D1325652

LIVERPOOL UNIVERSITY LIBRARY

WITHDRAWN
FROM
STOCK

CAMBRIDGE MONOGRAPHS ON MATHEMATICAL PHYSICS

General editors: P. V. Landshoff, D. R. Nelson, D. W. Sciama, S. Weinberg

HAMILTONIAN SYSTEMS
Chaos and quantization

HAMILTONIAN SYSTEMS
Chaos and quantization

ALFREDO M. OZORIO DE ALMEIDA

Instituto de Fisica, Universidade Estadual de Campinas

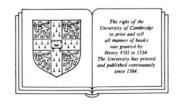

The right of the
University of Cambridge
to print and sell
all manner of books
was granted by
Henry VIII in 1534.
The University has printed
and published continuously
since 1584.

CAMBRIDGE UNIVERSITY PRESS

Cambridge

New York   New Rochelle   Melbourne   Sydney

Published by the Press Syndicate of the University of Cambridge
The Pitt Building, Trumpington Street, Cambridge CB2 1RP
32 East 57th Street, New York, NY 10022, USA
10 Stamford Road, Oakleigh, Melbourne 3166, Australia

© Cambridge University Press 1988

First published 1988

Printed in the United States of America

Library of Congress Cataloging-in-Publication Data
Ozorio de Almeida, Alfredo M.
Hamiltonian systems:chaos and quantization/Alfredo M. Ozorio
de Almeida.
p.   cm. – (Cambridge monographs on mathematical physics)
Translated from Portuguese.
Bibliography: p.
Includes index.
ISBN 0-521-34531-6
1. Hamiltonian systems.   2. Chaotic behavior in systems.
3. Quantum theory.   I. Title.   II. Series.
QC174.85.H35A45   1988   88-17039
530.1′3–dc 19   CIP

British Library Cataloguing in Publication Data
Ozorio de Almeida, Alfredo M.
Hamiltonian systems:chaos and
quantization. – (Cambridge monographs on
mathematical physics).
1. Dynamical systems
I. Title
003

ISBN 0 521 34531 6

Contents

Preface

The discovery of chaotic behaviour in deterministic dynamical systems has had a profound effect in many areas of physics. There is now a large literature on this subject, which includes some important surveys. However, a physicist just entering this beautiful field or an advanced graduate student still finds the need for a concise introduction to the basic concepts in unsophisticated mathematical language. What are the essential distinctions between Hamiltonian (conservative) systems and dissipative systems? In what way will the presence of chaos in the classical limit affect a system appropriately described by quantum mechanics?

There is much to be gained by studying the theory of Hamiltonian systems against the background of general systems, in order to emphasize both contrasts and similarities. Liouville's theorem provides a remarkable distinguishing feature of Hamiltonian systems. The preservation of volume in phase space prevents the asymptotic collapse of the motion onto equilibria, periodic orbits, or 'strange attractors'. Even though period-doubling cascades do occur in Hamiltonian systems, the loss of stability of a periodic orbit is only a local occurrence, rather than the·apocalyptic event that can subvert an entire dissipative system. In contrast the motion generated by a given Hamiltonian may exhibit diverse chaotic and regular orbits interwoven into rich structures.

There remains much to be unravelled concerning these global patterns, so that most of the classical theory in the first six chapters (Part I) deals with local motion. Periodic orbits, the fixed points of a Poincaré map, provide the unifying theme. Thus, we begin with the linearization around a fixed point and proceed to consider its structural stability, that is, the qualitative effect of a nonlinear perturbation. Chaotic motion appears as the natural consequence of homoclinic intersections of the separatrices emanating from unstable fixed points. We then apply some of the basic concepts of ergodic theory to the study of the proliferation of periodic orbits in chaotic systems. Infinite families of fixed points are calculated analytically as an application of the theory of normal forms. In chapter 6 the focus turns to the invariant tori of integrable systems and the delicate problem of their preservation under perturbations – KAM theory.

Part II is devoted to the semiclassical limit of quantum mechanics. In principle the entire scenario lying between integrability and chaos, which has been uncovered within classical dynamics, must be approximately mimicked by the corresponding quantized systems. Nevertheless, the well-established results presented in chapter 7 apply only to integrable systems. The wonderful self-consistency of this theory, originating in the old quantum mechanics, becomes in itself a barrier for its generalization. Therefore, the quantization of chaotic systems is not yet fully understood, but we discuss some important results obtained on the basis of the Weyl–Wigner formalism. We also study the computational evidence that the fluctuations of the energy spectra of chaotic systems coincide with those of random matrices. Finally we return to the classical periodic orbits – we explain their remarkable effect on the eigenstates and the spectrum of the Hamiltonian by making full use of the concepts developed in Part I.

Most of the fundamental work on classical dynamical systems has been carried out by mathematicians. Thus, the original literature compounds the difficulty of unfamiliar ideas (originating in topology, number theory, etc.) with the problem of translating modern mathematical language into the old-fashioned dialect still used by most physicists. The danger is that a complete translation would render the reader unfit to follow either the original references or future developments. I hope that, by providing minimal informal presentation of such concepts as diffeomorphisms and structural stability, this pitfall has been avoided. References are given only to the secondary literature from which most classical results have been obtained. Demonstrations present a further problem. Their inclusion was based on a double criterion, that is, simplicity and 'plausibility'. Thus, intuitive results are deduced only if the proof is devoid of technicalities. Most proofs are only sketched and the 'theorems' arise informally. The exercises scattered throughout the text range from simple applications that help the reader to assimilate the subject, to considerable extensions of the theory.

This book originated in a course of lectures given at the Universidade Estadual de Campinas during 1985. An earlier version was published by Editora da Unicamp in 1987, so the present text represents both a translation and a revision of the original Portuguese book. My general view of dynamical systems developed through years of collaboration with M. V. Berry and J. H. Hannay in Bristol. Professor Berry's review articles were extensively used, and I am also indebted to him for his careful scrutiny of the manuscript, though of course all its faults are my personal responsibility. Finally, I wish to thank Rosa Y. Kawaguchi for her precise typing, as well as Mary Nevader and my wife Susan for detecting and correcting grammatical anomalies in the English text.

Part I
Classical dynamics

1

Linear dynamical systems

Linear dynamical systems and their discrete counterparts, linear maps, have complete explicit solutions. In the former case these are the basis for approximating the motion near points of equilibrium of nonlinear systems. A similar theory for maps elucidates the motion surrounding periodic orbits, as a consequence of the Poincaré section, to be introduced in chapter 2. We shall first review the motion of nonconservative systems and then consider the restrictions peculiar to Hamiltonian systems. The importance of the linear flow on a torus, discussed in section 1.3, will manifest itself fully only through the study of integrable systems in chapter 6, but it provides a ready example of the concept of ergodicity, to be developed in chapter 3. Even more important, as a point of departure for the study of chaotic motion, is the study of linear maps of the torus in section 1.4. In spite of the elementary nature of the mathematical techniques employed, it will thus be found that this initial chapter is the germ for the whole of Part I.

A *dynamical system* is defined as a set of N first-order differential equations in N variables $\mathbf{x} = (x_1, \ldots, x_N)$ (these are said to determine a point in *phase space*, the Euclidean space $\mathbb{R}^N$, unless otherwise stated):

$$dx/dt \equiv \dot{\mathbf{x}} = \mathbf{f}(\mathbf{x}, t). \qquad (1.1)$$

The system will be *autonomous* if the *vector field* $\mathbf{f}$ is independent of t, the time. A *solution* of the dynamical system is a vector function $\mathbf{x}(\mathbf{x}_0, t)$, satisfying (1.1) and the initial condition

$$\mathbf{x}(\mathbf{x}_0, 0) = \mathbf{x}_0. \qquad (1.2)$$

Often we omit the dependence on the initial condition and refer to the solution simply as $\mathbf{x}(t)$.

An important class of solutions is that of the *points of equilibrium* $\mathbf{x}_e$ for which $\mathbf{f}(\mathbf{x}, t) = 0$. The expression of the vector field in a Taylor series,

$$\mathbf{f}(\mathbf{x}, t) = \frac{\partial \mathbf{f}(\mathbf{x}_e, t)}{\partial \mathbf{x}}(\mathbf{x} - \mathbf{x}_e) + \frac{1}{2}(\mathbf{x} - \mathbf{x}_e)\frac{\partial^2 \mathbf{f}(\mathbf{x}_e, t)}{\partial \mathbf{x}^2}(\mathbf{x} - \mathbf{x}_e) + \cdots \qquad (1.3)$$

(where $\partial \mathbf{f}/\partial \mathbf{x}$ is the Jacobian matrix), can be truncated so as to yield a good approximation of the motion near equilibrium, even if the Taylor series does not converge. For an autonomous field the first approximation to the dynamical system is brought into the form

$$\dot{\mathbf{x}} = \mathscr{A}\mathbf{x} \tag{1.4}$$

by changing the origin of coordinates to $\mathbf{x}_e$ and defining the *dynamical matrix* $\mathscr{A} = \partial \mathbf{f}(0)/\partial \mathbf{x}$. In the remainder of this chapter we will study the solutions of the linear system (1.4), without any reference to the possible nonlinear systems that it may approximate.

1.1 General linear systems

The solution of the linear system (1.4) is simply

$$\mathbf{x}(\mathbf{x}_0, t) = \exp(t\mathscr{A})\mathbf{x}_0, \tag{1.5}$$

as follows from the series expansion of the exponential matrix

$$\exp(t\mathscr{A}) = 1 + t\mathscr{A} + \frac{t^2}{2}\mathscr{A}^2 + \cdots + \frac{t^n}{n!}\mathscr{A}^n + \cdots. \tag{1.6}$$

The *general solution* of the linear system is a linear superposition of N linearly independent solutions $\{\mathbf{x}^1(t), \ldots, \mathbf{x}^N(t)\}$:

$$\mathbf{x}(t) = \sum_{n=1}^{N} C_n \mathbf{x}^n(t), \tag{1.7}$$

where the coefficients C_n are determined by the initial condition.

The nature of the flow around the equilibrium at the origin depends exclusively on the eigenvalues $\{\lambda_1, \ldots, \lambda_N\}$ of the dynamical matrix. If all the eigenvalues are real and nondegenerate, the matrix can be diagonalized by a real linear transformation. This will also diagonalize the solution matrix $\exp(t\mathscr{A})$, the eigenvalues of which are $\{\exp(t\lambda_1), \ldots, \exp(t\lambda_N)\}$. Given $\boldsymbol{\xi}_n$, the eigenvectors of $\mathscr{A}$, we then have N independent solutions:

$$\mathbf{x}^n(t) = \exp(\lambda_n t)\boldsymbol{\xi}_n. \tag{1.8}$$

A simple example from Arnold (1973) helps to bring home the idea: Let

$$\mathscr{A} = \begin{bmatrix} 1 & 1 \\ 1 & 1 \end{bmatrix}$$

in an arbitrary basis $(\mathbf{e}_1, \mathbf{e}_2)$. The eigenvalues are then $\{2, 0\}$, with eigenvectors $\boldsymbol{\xi}_1 = \mathbf{e}_1 + \mathbf{e}_2$ and $\boldsymbol{\xi}_2 = \mathbf{e}_1 - \mathbf{e}_2$. In the basis of $(\boldsymbol{\xi}_1, \boldsymbol{\xi}_2)$ the solution matrix is just

$$\begin{bmatrix} \exp(2t) & 0 \\ 0 & 1 \end{bmatrix},$$

so the general solution is

$$\mathbf{x}(t) = c_1(1, 1)\exp(2t) + c_2(1, -1)$$
$$= (c_1 \exp(2t) + c_2, c_1 \exp(2t) - c_2).$$

If there are any complex eigenvalues, the dynamical matrix can be diagonalized only by a complex linear transformation. It is thus useful to *complexify* the dynamical system. This procedure is also valid for the general system (1.1), which takes the form

$$\dot{\mathbf{z}} = \mathbf{f}(\mathbf{z}, t), \tag{1.9}$$

where $\mathbf{z}$ belongs to $\mathbf{C}^N$, a space of N complex variables. So if the vectors $\{\mathbf{e}_1, \ldots, \mathbf{e}_n\}$ form a basis in the real space $\mathbb{R}^N$, the vectors $\{\mathbf{e}_1 + i\mathbf{0}, \ldots, \mathbf{e}_N + i\mathbf{0}\}$ form a basis in $\mathbf{C}^N$. The expansion of an arbitrary complex vector,

$$\boldsymbol{\zeta} = \sum_{n=1}^{N} z_n \mathbf{e}_n, \tag{1.10}$$

is then obtained from the rule for the multiplication of a vector by a scalar:

$$(u + iv)(\boldsymbol{\xi} + i\boldsymbol{\eta}) = (u\boldsymbol{\xi} - v\boldsymbol{\eta}) + i(v\boldsymbol{\xi} + u\boldsymbol{\eta}). \tag{1.11}$$

The vector $\boldsymbol{\xi} + i\mathbf{0}$ will hereafter be abbreviated $\boldsymbol{\xi}$.

The vector field $\mathbf{f}$ is still a set of real functions, so that its *complex conjugate field* $\mathbf{f}^*$ is given by

$$\mathbf{f}^*(\mathbf{z}, t) = \mathbf{f}(\mathbf{z}^*, t). \tag{1.12}$$

This leads to the fundamental property of the complexified system: If $\mathbf{z}(\mathbf{z}_0, t)$ is a solution, then $\mathbf{z}^*(\mathbf{z}_0, t)$ is also a solution, as shown in fig. 1.1. If $\mathbf{z}_0$ is real, the uniqueness theorem for differential equations entails that $\mathbf{z}(\mathbf{z}_0, t)$ is real for all time. In the specific case of the linear system

$$\dot{\mathbf{z}} = \mathscr{A}\mathbf{z}, \tag{1.13}$$

with $\mathscr{A}$ a real matrix, the equality

$$\mathscr{A}(\mathbf{x} + i\mathbf{y}) = \mathscr{A}\mathbf{x} + i\mathscr{A}\mathbf{y} \tag{1.14}$$

implies that, if $\mathbf{z}(\mathbf{z}_0, t)$ is a solution of the system (1.13), then the real and imaginary parts are solutions of the real system (1.4). Therefore, we can find the real solutions from the complex solutions and vice versa; that is,

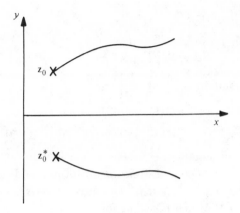

Fig. 1.1. The vector field $\mathbf{f}$ is real; then $\mathbf{z}(\mathbf{z}_0, t)$, being a complex solution, implies that $\mathbf{z}^*(\mathbf{z}_0, t)$ is also a solution.

we have

$$\mathbf{z}(\mathbf{z}_0, t) = \mathbf{x}(\mathbf{x}_0, t) + i\mathbf{y}(\mathbf{y}_0, t). \tag{1.15}$$

The reader should consult Arnold (1973) for a fuller discussion of complexification.

The solutions of the complex system (1.13) are of the same form as the real solutions. Therefore, there is a complex solution $\exp(\lambda_n t)\zeta_n$ for every eigenvalue λ_n, whether real or complex, and the general solution is

$$\mathbf{z}(t) = \sum_{n=1}^{N} C_n \exp(\lambda_n t)\zeta_n. \tag{1.16}$$

The complex conjugate of this expression is also a general solution,

$$\mathbf{z}^*(t) = \sum_{n=1}^{N} C_n^* \exp(\lambda_n^* t)\zeta_n^*, \tag{1.17}$$

so complex eigenvalues must come in pairs λ_n, λ_n^* with corresponding eigenvectors ζ_n, ζ_n^*. We conclude that, if the number of real eigenvalues is N_{R} and the number of pairs of complex eigenvectors is N_{C} ($N = N_{\mathrm{R}} + 2N_{\mathrm{C}}$), the general solution of a real system with distinct eigenvalues is

$$\mathbf{x}(t) = \sum_{n=1}^{N_{\mathrm{R}}} a_n \exp(\lambda_n t)\xi_n + \sum_{n=1}^{N_{\mathrm{C}}} \mathrm{Re}\{C_n \exp(\lambda_n t)\zeta_n\}, \tag{1.18}$$

where $\mathrm{Re}\{\ \}$ denotes the real part of a complex number. Each real eigenvector determines motion along a line, whereas complex eigenvectors give rise to motion in a plane.

Exercise

Decomposing $\zeta_j = \xi_j + i\eta_j$ and $\lambda_j = \alpha_j + i\omega_j$, show that

$$\xi_j(t) = e^{\alpha_j t}[\cos(\omega_j t)\xi_j - \sin(\omega_j t)\eta_j],$$
$$\eta_j(t) = e^{\alpha_j t}[\sin(\omega_j t)\xi_j + \cos(\omega_j t)\eta_j],$$
(1.19)

so that, using ξ_j and η_j as basis vectors, the projection of the initial condition onto the (ξ_j, η_j) plane $x_j\xi_j + y_j\eta_j$ is given by

$$x_j(t) = e^{\alpha_j t}[\cos(\omega_j t)x_j(0) + \sin(\omega_j t)y_j(0)],$$
$$y_j(t) = e^{\alpha_j t}[-\sin(\omega_j t)x_j(0) + \cos(\omega_j t)y_j(0)].$$
(1.20)

This is the product of a contraction $e^{\alpha_j t}$ (or expansion if $\alpha_j > 0$) composed with a rotation by the angle $\omega_j t$ (fig. 1.2).

The origin is a stable or unstable *focus* in the plane of motion determined by a general complex eigenvalue. If the eigenvalue is purely imaginary, the origin will be a *centre* surrounded by circles in the eigencoordinates, or ellipses in general coordinates. In the case of real eigenvalues the orbits are

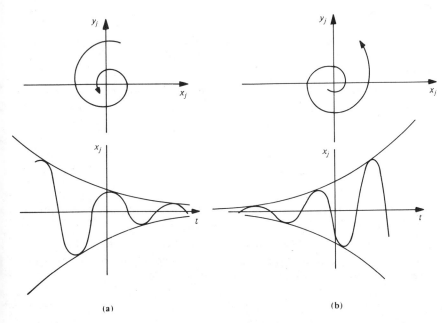

(a) (b)

Fig. 1.2. Linear motion in the plane with a general complex eigenvalue is the product of an exponential contraction or expansion composed with a rotation. (a) $\alpha < 0$; $\omega < 0$. (b) $\alpha > 0$; $\omega < 0$.

given by

$$x_2 = cx_1^{\lambda_2/\lambda_1}. \tag{1.21}$$

If $\lambda_2/\lambda_1 < 0$, the origin will be a *saddle point*, whereas $\lambda_1 < \lambda_2 < 0$ determines a stable *node* and $0 < \lambda_1 < \lambda_2$ an unstable node. All these possibilities appear in fig. 1.3.

Eigenvalue degeneracies may be an obstacle to the diagonalization of the dynamical matrix. This is not a problem for the hermitian or unitary matrices that commonly turn up in physics, but the matrices arising in dynamical systems can be quite general. The crucial factor is the rank of the matrix $\{\mathscr{A} - \lambda\mathbf{1}\}$, where λ is one of the eigenvalues. If λ is non-degenerate, rank $\{\mathscr{A} - \lambda\mathbf{1}\} = N - 1$, and there exists an eigenvector corresponding to λ. If λ is μ-fold degenerate and rank $\{\mathscr{A} - \lambda\mathbf{1}\} = N - \mu$, λ will be the eigenvalue of μ linearly independent eigenvectors. However, if rank $\{\mathscr{A} - \lambda\mathbf{1}\} = N - \nu$ where $\nu < \mu$, there will be only ν eigenvectors corresponding to λ. In this case the matrix cannot be diagonalized.

Example. The matrix

$$\mathscr{A} = \begin{bmatrix} 1 & 1 \\ 0 & 1 \end{bmatrix},$$

for a horizontal shear in the plane, has the doubly degenerate eigenvalue 1. Rank $\{\mathscr{A} - \mathbf{1}\} = 1$, so $\mathscr{A}$ cannot be diagonalized. Its unique eigenvector is $(1, 0)$.

In general the best that we can do is to reduce the dynamical matrix to its *normal form*, where each eigenvector corresponds to a *Jordan block*:

$$\mathscr{B} = \begin{bmatrix} \lambda & 1 & & & 0 \\ & \lambda & 1 & & \\ & & \ddots & \ddots & \\ 0 & & & \lambda & 1 \end{bmatrix}. \tag{1.22}$$

The sum of the dimensions of all the Jordan blocks corresponding to a given eigenvalue will be μ, the order of its degeneracy.

The dynamics of a system with degenerate eigenvalues can be reduced by a linear transformation to the dynamics of individual blocks. To calculate $\exp(\mathscr{B}t)$ we follow Arnold (1973) in writing $\mathscr{B} = \lambda\mathbf{1} + \mathscr{I}$ with

$$\mathscr{I} = \begin{bmatrix} 0 & 1 & & & 0 \\ & 0 & 1 & & \\ & & \ddots & \ddots & \\ & & & & 1 \\ 0 & & & & 0 \end{bmatrix}. \tag{1.23}$$

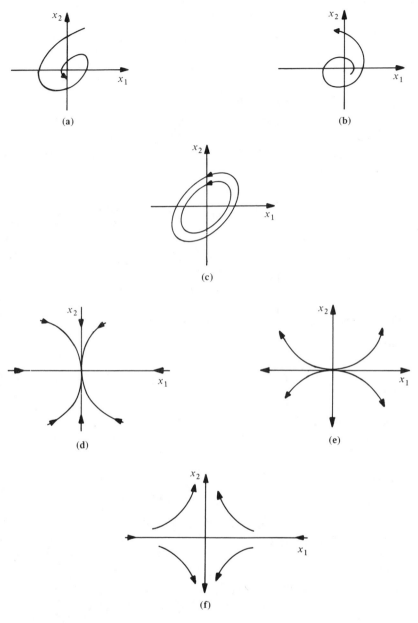

Fig. 1.3. Typical linear motions in a plane. (a) Stable focus; (b) unstable focus; (c) centre; (d) stable node; (e) unstable node; (f) saddle.

The matrix $\exp(\mathscr{S}t)$ is easily calculated by observing that $\mathscr{S}$ operates on the basis vectors $\{\mathbf{e}_1,\ldots,\mathbf{e}_N\}$ through the simple shift $0\leftarrow\mathbf{e}_1\leftarrow\mathbf{e}_2\cdots\leftarrow\mathbf{e}_n$. Therefore, $\mathscr{S}^k$ operates the shift $\mathbf{e}_{m-k}\leftarrow\mathbf{e}_m\leftarrow\mathbf{e}_{m+k}$; that is,

$$\mathscr{S}^k = \begin{bmatrix} 0 & \cdots & 1 & & 0 \\ & & & \ddots & \\ & & & & 1 \\ & & & & \vdots \\ 0 & & & & 0 \end{bmatrix} \tag{1.24}$$

and $\mathscr{S}^N = 0$. The series representation of $\exp(\mathscr{S}t)$ thus leads immediately to

$$e^{\mathscr{S}t} = \begin{bmatrix} 1 & t & t^2/2 & \cdots & t^{n-1}/(n-1)! \\ & \ddots & \ddots & \ddots & \vdots \\ & & & & t^2/2 \\ & & & & t \\ 0 & & & & 1 \end{bmatrix}. \tag{1.25}$$

Exercise

Show that if two matrices $\mathscr{A}$ and $\mathscr{B}$ commute, then

$$e^{\mathscr{A}+\mathscr{B}} = e^{\mathscr{A}}e^{\mathscr{B}}. \tag{1.26}$$

Combining (1.25) and (1.26), we conclude that the evolution of a Jordan block is given by

$$e^{\mathscr{B}t} = \begin{bmatrix} e^{\lambda t} & te^{\lambda t} & \cdots & t^{n-1}e^{\lambda t}/(n-1)! \\ & e^{\lambda t} & \ddots & \vdots \\ & & \ddots & te^{\lambda t} \\ 0 & & & e^{\lambda t} \end{bmatrix}. \tag{1.27}$$

Example (Arnold, 1973). Consider the dynamical system obtained from the Nth-order differential equation

$$x^{(N)} = a_1 x^{(N-1)} + \cdots + a_{N-1}\dot{x} + a_N x, \tag{1.28}$$

through the identification of each derivative with a coordinate: $x^{(n)} = d^n x/dt^n = x_n$. This linear system has the dynamical matrix

$$\mathscr{A} = \begin{bmatrix} 0 & 1 & & 0 \\ & & \ddots & \ddots \\ & & 0 & 1 \\ a_n & \cdots & a_2 & a_1 \end{bmatrix}. \tag{1.29}$$

The trial solution $\exp(\lambda t)x_0$ for equation (1.29) leads to the algebraic equation

$$\lambda^N = a_1 \lambda^{N-1} + \cdots + a_N, \tag{1.30}$$

with solutions $\lambda_1, \ldots, \lambda_k$, with $k \leq N$. No matter how degenerate the eigenvalue λ may be, it will correspond to a unique eigenvector of (1.29):

$$\left(e^{\lambda t}x_0, \ldots, \frac{d^{N-1}}{dt^{N-1}} e^{\lambda t}x_0 \right)_{t=0} = (x_0, \ldots, \lambda^{N-1}x_0). \tag{1.31}$$

Thus, in this system each degenerate eigenvalue will correspond to a single Jordan block.

In this section we have gone through all the ways in which a linear system can be reduced to a set of simpler linear systems. To conclude, we note that the *flow* of a linear system $x(x_0, t)$ defines a linear map $x_0 \to x$ for all t, according to (1.5). Conversely, any *linear map*

$$x' = \mathscr{F}x \tag{1.32}$$

with positive eigenvalues can be interpreted as the flow after a unit of time of an autonomous dynamical system defined by the dynamical matrix $\mathscr{A} = \ln \mathscr{F}$. The result of following the flow for k units of time is the same as that of k iterations of the map. General flows always define maps, but the reciprocal equivalence rarely holds. It does not apply to many nonlinear maps obtained as 'Poincaré sections' in chapter 2. In particular these may have negative eigenvalues, denoting reflection of the motion across the origin in the direction of the eigenvector. In this case it is only the square of the linearized map that can be fitted by a real flow.

1.2 Linear Hamiltonian systems

Hamiltonian systems are characterized by a *Hamiltonian function* $H(x, t) = H(p, q, t)$ of an even number of variables $p = (p_1, \ldots, p_L)$ and $q = (q_1, \ldots, q_L)$, where L is known as the *number of freedoms*. The dynamical system is defined by *Hamilton's equations*:

$$\dot{p} = -\frac{\partial H}{\partial q}, \qquad \dot{q} = \frac{\partial H}{\partial p}. \tag{1.33}$$

If the Hamiltonian is independent of time, the system is *autonomous* and $H(p, q)$ is a *constant* or *integral* of the motion:

$$\frac{dH}{dt} = \frac{\partial H}{\partial p} \dot{p} + \frac{\partial H}{\partial q} \dot{q} = 0. \tag{1.34}$$

The system (1.33) can be linearized, which is tantamount to approximating the Hamiltonian by the quadratic form $(\mathbf{x}\mathcal{H}\mathbf{x})/2$, where $\mathcal{H}$ is the *Hessian matrix*

$$\mathcal{H} = \frac{\partial^2 H}{\partial \mathbf{x} \partial \mathbf{x}}. \tag{1.35}$$

A linear Hamiltonian system thus has the form

$$\dot{\mathbf{x}} = \mathcal{J}\mathcal{H}\mathbf{x}, \tag{1.36}$$

where $\mathcal{J}$ is the $(2L \times 2L)$-dimensional matrix

$$\mathcal{J} = \left[\begin{array}{c|c} \mathbf{0} & -\mathbf{1} \\ \hline \mathbf{1} & \mathbf{0} \end{array}\right]. \tag{1.37}$$

Defining $\mathcal{J}'$ as the transposed matrix, we verify immediately that $\mathcal{J}' = \mathcal{J}^{-1} = -\mathcal{J}$. The product $\mathcal{J}\mathcal{H}$ will be referred to as the *Hamiltonian matrix*.

The *symplectic area* of the parallelogram defined by two vectors $\boldsymbol{\xi} = (\boldsymbol{\xi}_p, \boldsymbol{\xi}_q)$ and $\boldsymbol{\eta} = (\boldsymbol{\eta}_p, \boldsymbol{\eta}_q)$ is given by the *skew product*

$$\boldsymbol{\xi} \wedge \boldsymbol{\eta} = (\mathcal{J}\boldsymbol{\xi}) \cdot \boldsymbol{\eta} = \sum_{k=1}^{L} (\xi_{pk}\eta_{qk} - \xi_{qk}\eta_{pk}), \tag{1.38}$$

where the dot indicates the ordinary scalar product. (See Arnold, 1978, for a complete discussion of skew products and *exterior algebra*.) Geometrically this measures the algebraic sum of the areas of the parallelograms formed by the projections of $\boldsymbol{\xi}$ and $\boldsymbol{\eta}$ onto the L *conjugate planes* with coordinates p_k and q_k (fig. 1.4). The symplectic area is conserved by a linear Hamiltonian

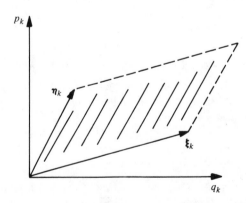

Fig. 1.4. The symplectic area of a closed curve in phase space is the algebraic sum of the areas of its projections onto all the canonical conjugate planes. In the case of the parallelogram formed by two vectors $\boldsymbol{\xi}$ and $\boldsymbol{\eta}$, this is the skew product $\boldsymbol{\xi} \wedge \boldsymbol{\eta}$.

flow, since

$$\frac{d}{dt}(\xi \wedge \eta) = (\mathscr{J}\dot{\xi}) \cdot \eta + (\mathscr{J}\xi) \cdot \dot{\eta}$$
$$= (\mathscr{J}\mathscr{J}\mathscr{H}\xi) \cdot \eta + (\mathscr{J}\xi) \cdot (\mathscr{J}\mathscr{H}\eta) = 0. \tag{1.39}$$

Exercises

1. Prove that $\xi \wedge \eta = -\eta \wedge \xi$ and hence $\xi \wedge \xi = 0$.
2. Show that any vector ξ defines a unique $(2L-1)$-dimensional plane, which is *skew-orthogonal* to ξ, that is, $\xi \wedge \eta = 0$ for all vectors η in the plane. Use the similar results for the scalar product, but note that the plane includes ξ because of exercise 1.

Area invariance is a fundamental result for Hamiltonian systems. For a start it provides a purely algebraic condition for the matrix $\mathscr{M} = \exp(\mathscr{A}t)$ to describe a Hamiltonian flow: We must have

$$(\mathscr{J}\mathscr{M}\xi) \cdot (\mathscr{M}\eta) = (\mathscr{J}\xi) \cdot \eta, \tag{1.40}$$

which reduces to

$$\mathscr{M}'\mathscr{J}\mathscr{M} = \mathscr{J}, \tag{1.41}$$

the definition of a *symplectic matrix*.

For a nonlinear system, $\mathscr{M}$ is the Jacobian matrix of the full *canonical transformation* generated by the Hamiltonian flow, so that (1.41) must hold for all points in the phase space. Taking the determinant of both sides of (1.41) yields

$$\det(\mathscr{M}) = \pm 1. \tag{1.42}$$

If $\mathscr{M} = \exp(\mathscr{J}\mathscr{H}t)$, $\det(\mathscr{M})$ will be continuous in t, so from the initial condition at $t = 0$, we have

$$\det(\mathscr{M}) = 1 \tag{1.43}$$

for a Hamiltonian flow. This is the differential form of *Liouville's theorem*.

Exercise

A *point transformation* does not mix the **p** and the **q** coordinates. Its matrix therefore has the form

$$\left[\begin{array}{c|c} \mathscr{A} & \mathbf{0} \\ \hline \mathbf{0} & \mathscr{B} \end{array}\right].$$

Show that the condition for this to be symplectic is that $\mathscr{A}\mathscr{B}' = \mathscr{A}'\mathscr{B} = \mathbf{1}$. Thus, $\mathscr{B} = \mathscr{A}$ if $\mathscr{A}$ is unitary.

Next we deduce that, if $\gamma \neq 0$ is an eigenvalue of a symplectic matrix $\mathscr{M}$, then γ^{-1} is also an eigenvalue. The characteristic polynomial $P(\gamma)$ can be written in the forms

$$P(\gamma) = \det[\mathscr{M} - \gamma\mathbf{1}] = \det[\mathscr{J}^{-1}\mathscr{M}'^{-1}\mathscr{J} - \gamma\mathbf{1}] = \det[\mathscr{M}'^{-1} - \gamma\mathbf{1}]$$

$$= \frac{(-\gamma)^{2L}}{\det(\mathscr{M}')}\det[\mathscr{M}' - \gamma^{-1}\mathbf{1}] = \frac{\gamma^{2L}}{\det(\mathscr{M})}\det[\mathscr{M} - \gamma^{-1}\mathbf{1}]. \qquad (1.44)$$

Thus,

$$P(\gamma) = \pm\gamma^{2L}P(\gamma^{-1}), \qquad (1.45)$$

and if γ_0 is a root of $P(\gamma)$, so is γ_0^{-1}. Incidentally this shows that the sign is never negative in (1.42), even if $\mathscr{M}$ has negative eigenvalues.

The implication is that the nonzero eigenvalues of the Hamiltonian matrix $\mathscr{J}\mathscr{H}$ always come in pairs $(\lambda, -\lambda)$. Combining the condition that $\mathscr{H}$ is real, which imposes complex (λ, λ^*) as simultaneous pairs, we elicit the four types of eigenvalue for the Hamiltonian matrix:

1. Zero eigenvalues
2. Pairs of real eigenvalues $\pm\lambda$
3. Pairs of purely imaginary eigenvalues $\pm i\omega$
4. Quartets of complex eigenvalues $\pm\alpha \pm i\omega$

A linear Hamiltonian system with only one freedom must fall into one of the first three categories (though category 1 is the trivial system in this case). Their presence is also inevitable if the number of freedoms is odd.

It may seem that any excess of real or imaginary eigenvalues will move into the complex plane after a small general quadratic perturbation of the Hamiltonian. However, this requires the simultaneous exit of four eigenvalues from the real or the imaginary axis, so it is only when two pairs of eigenvalues in class (ii) or (iii) degenerate that they can form a complex quartet. In a one-parameter family of systems this event, illustrated in fig. 1.5, will occur at isolated values of the parameter. The necessary and sufficient conditions are provided in Arnold (1978, sec. 4.2). Thus, real and imaginary eigenvalues are stable in the family of Hamiltonian systems. Another way to see this is that two pairs of eigenvalues on an axis have 'two freedoms' in which to move, just as a single quartet in the plane. Only the zero eigenvalue has zero probability of showing up in a randomly picked quadratic Hamiltonian. A property that is usually resistant to a given class

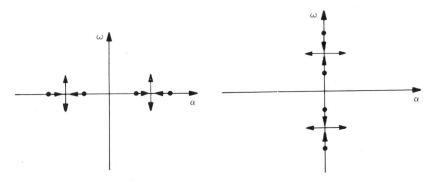

Fig. 1.5. Real eigenvalues and imaginary eigenvalues are structurally stable, because it is required that two pairs of eigenvalues leave the axis simultaneously.

of perturbations of the system is described as *generic*. The system with generic properties is itself called generic or 'structurally stable'.

Generic Hamiltonian matrices do not have eigenvalue degeneracies and can therefore be diagonalized by a linear transformation. The condition for the transformed system to still be Hamiltonian is that the transformation be symplectic. Thus, in the new system we must have the skew products among the basis vectors

$$\hat{\mathbf{p}}_i \wedge \hat{\mathbf{q}}_j = \delta_{ij}; \qquad \hat{\mathbf{p}}_i \wedge \hat{\mathbf{p}}_j = \hat{\mathbf{q}}_i \wedge \hat{\mathbf{q}}_j = 0, \tag{1.46}$$

just as in the old system. This condition holds for the eigenvectors of the Hamiltonian matrix. In the case of a pair of real eigenvalues λ_1 and λ_2, the symplectic area of the parallelogram defined by the corresponding eigenvectors ξ_1, ξ_2 would be multiplied by $\exp[(\lambda_1 + \lambda_2)t]$ after a time t if it were not zero. So we can take these eigenvectors as the new axes $\hat{\mathbf{p}}_1$ and $\hat{\mathbf{p}}_2$. On the other hand, symplectic areas in the plane of ξ_1 and ξ_{-1} corresponding to the eigenvalues $\pm \lambda_1$ are conserved, enabling us to choose ξ_{-1} as the new $\hat{\mathbf{q}}_1$ axis, apart from normalization. In the case of imaginary eigenvalues ω_j, the new $(\hat{\mathbf{p}}_j, \hat{\mathbf{q}}_j)$ plane is just the $(\xi_j, \boldsymbol{\eta}_j)$, corresponding to the real and imaginary part of the eigenvector ζ_j, as in (1.19).

Exercise

Show that the choice of basis vectors $(\hat{p}_1, \hat{p}_2) = (\xi, \boldsymbol{\eta})$ in (1.19) for the eigenvalue $\alpha + i\omega$ and $(\hat{q}_1, \hat{q}_2) = (\xi', \boldsymbol{\eta}')$ for the eigenvalue $-\alpha + i\omega$ implies the invariance of the symplectic area of any parallelogram in the $(\hat{p}_1, \hat{p}_2, \hat{q}_1, \hat{q}_2)$ subspace.

The *normal forms* for the quadratic Hamiltonians that generate these independent submatrices are of three types:

1. Real eigenvalues $(\pm \lambda)$: $H = -\lambda p_1 q_1$
2. Imaginary eigenvalues $(\pm i\omega)$: $H = \pm (\omega/2)(p_1^2 + q_1^2)$
3. Complex eigenvalues $(\pm \alpha \pm i\omega)$: $H = -\alpha(p_1 q_1 + p_2 q_2) + \omega(p_1 q_2 - p_2 q_1)$

Generic quadratic Hamiltonians can be reduced to a linear superposition of these simple forms.

Exercises

1. Show that the symplectic transformation, which rotates each pair of conjugate axes by $\pi/4$, takes the Hamiltonian of type 3 into the form

$$H = -\frac{\alpha}{2}(p_1^2 - q_1^2 + p_2^2 - q_2^2) + \omega(p_1 q_2 - p_2 q_1). \tag{1.47}$$

2. Show that (1.47) describes the quadratic approximation for the Hamiltonian of a spherical pendulum near the unstable equilibrium, in coordinates that rotate around the vertical axis. (This interpretation was communicated to me by J. Koiller.)

Isolated degeneracies may occur in a typical one-parameter family of Hamiltonians. Some of the Jordan blocks that result are *generic*, that is, *structurally stable* with respect to small variations of the *family*. The normal form of the Hamiltonian for each of these blocks is given in Arnold (1978, app. 6) as

4. Zero eigenvalues (0): $H = \pm \frac{1}{2} q_1^2$
5. Real eigenvalues $(\pm \lambda)$: $H = -\lambda(p_1 q_1 + p_2 q_2) + p_1 q_2$
6. Imaginary eigenvalues $(\pm i\omega)$: $H = \pm \frac{1}{2}(q_1^2/\omega^2 + q_2^2) - \omega^2 p_1 q_2 + p_2 q_1$

The same reference also provides the generic Jordan blocks in two-parameter families.

1.3 Linear flow on a torus

All of the eigenvalues of a linear Hamiltonian system may well be imaginary. This is the only possibility that leads to stable motion near the origin, since the other alternatives imply the existence of a positive real part of at least one of the eigenvalues. Choosing the (p_j, q_j) coordinate axes along the real and imaginary parts of the L eigenvectors ζ_j, the equations of motion take the form

$$\dot{p}_j = -\omega_j q_j, \qquad \dot{q}_j = \omega_j p_j. \tag{1.48}$$

Transforming to *canonical polar coordinates* [*action-angle variables* $(\mathbf{I}, \boldsymbol{\theta})$],

$$p_j = (2I_j)^{1/2} \sin \theta_j, \qquad q_j = (2I_j)^{1/2} \cos \theta_j, \tag{1.49}$$

the system reduces to

$$\dot{I}_j = 0, \qquad \dot{\theta}_j = \omega_j. \tag{1.50}$$

For a system with a single freedom, $I = \text{const}$ defines a circle. In a two-freedom system $\mathbf{I} = \text{const}$ is the direct product of two circles, that is, a two-dimensional *torus* in the four-dimensional phase space. The two angular coordinates $\theta_j + 2\pi = \theta_j$ specify the 'latitude and longitude' on the torus, as shown in fig. 1.6a. In the (θ_1, θ_2) coordinate plane, each of the 2π-sided squares shown in fig. 1.6b is an image of the torus. The solution of the system is an orbit that winds itself around the torus, while in the (θ_1, θ_2) coordinates it is a straight line.

In the general case the L-dimensional surface, defined by $\mathbf{I} = \text{const}$, has L independent irreducible circuits corresponding to the displacement of 2π for each of the angles θ_j. This surface is *an L torus*. The orbits wind around the L torus, though they are parallel straight lines in the space of the $\boldsymbol{\theta}$ coordinates.

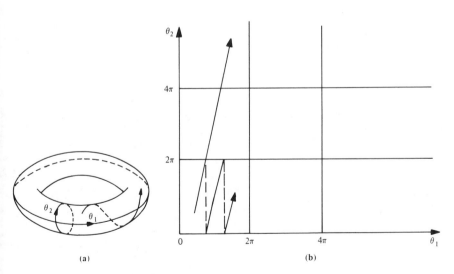

(a) (b)

Fig. 1.6. The orbits of (1.50), which wind around the torus (a), are straight lines in the angular coordinates (b).

A notable way to study motion on the torus is to consider the repeated traversals of an orbit through the 'meridian' $\theta_1 = 0 \pmod{2\pi}$, that is, the *Poincaré section* of the torus. The latitudes of the intersections for the solutions of (1.50) will be

$$\theta_{2k} = \theta_{20} + k\alpha, \tag{1.51}$$

where $\alpha = 2\pi\omega_2/\omega_1$ (fig. 1.7). The mapping $\theta_2 \to \theta_2' = \theta_2 + \alpha$ for all the points of the circle is known as the *Poincaré map*, a uniform translation in this instance.

There will be two distinct cases. If $\omega_2/\omega_1 = r/s$ with r and s mutually prime integers, then $\theta_2 s = \theta_{20}$. The orbits of successive iterations of the Poincaré map will be a set of s equally spaced points. An orbit of (1.50) draws out a finite set of parallel segments in a 2π-sided square. The alternative is that ω_2/ω_1 is irrational, in which case $\theta_{2k} \neq \theta_{20}$ for all k; that is, the orbit on the torus never closes.

It is easy to show that such an orbit is *dense* on the circle: Following Arnold (1973) we divide the circle into k equal semiopen intervals. Among the first $k + 1$ images of θ_{20}, at least two will share the same interval. Let these two points be $\theta_{20} + p\alpha$ and $\theta_{20} + q\alpha$, with $s = p - q > 0$. Thus,

$$|s\alpha|(\text{mod } 2\pi) < 2\pi/k, \tag{1.52}$$

implying that some point in the sequence $\theta_{20} + js\alpha$ comes within the

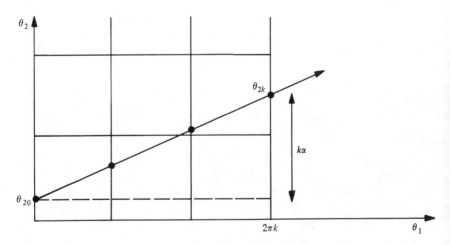

Fig. 1.7. Successive intersections of an orbit with the line $\theta_2 = 2\pi k$ define its orbit under the Poincaré map.

distance $\varepsilon = 2\pi/k$ of any point in the circle. Since ε may be chosen arbitrarily small, the orbit is dense.

It follows that the orbits for the linear motion on the torus are also dense. In the case of an L torus, this will be the case if the L frequencies are *rationally independent*, that is, on condition that

$$k_1\omega_1 + \cdots + k_L\omega_L = 0, \tag{1.53}$$

if and only if $k_1 = \cdots = k_L = 0$.

Let us now consider an arbitrary finite function $f(\boldsymbol{\theta})$ defined on the torus. A dense orbit will sample points 'democratically' in its way along the torus, so it is plausible that the time average

$$\bar{f}(\boldsymbol{\theta}_0) = \lim_{T \to \infty} \frac{1}{T} \int_0^T f(\boldsymbol{\theta}(\boldsymbol{\theta}_0, t))\, dt \tag{1.54}$$

exists and that it equals the spatial average:

$$\bar{f}(\boldsymbol{\theta}_0) = (2\pi)^{-L} \int f(\boldsymbol{\theta})\, d\boldsymbol{\theta}. \tag{1.55}$$

This equality is verified for harmonic functions $f(\boldsymbol{\theta}) = \exp(i\mathbf{k}\cdot\boldsymbol{\theta})$, where $\mathbf{k}$ is an integer vector, for rationally independent frequencies. We then have

$$\frac{1}{T} \int_0^T dt\, \exp(i\mathbf{k}\cdot(\boldsymbol{\theta}_0 + \boldsymbol{\omega}t))\, dt = \frac{\exp(i\mathbf{k}\cdot\boldsymbol{\theta}_0)}{i(\mathbf{k}\cdot\boldsymbol{\omega})T}[\exp(i\mathbf{k}\cdot\boldsymbol{\omega}T) - 1]. \tag{1.56}$$

The function in the square brackets is bounded, so the limit of the right side is zero if $\mathbf{k} \neq 0$, whereas (1.55) is trivially verified if $\mathbf{k} = 0$. Since both averages are linear, (1.55) is also valid for any trigonometric polynomial and hence for any truncation of the Fourier representation of an arbitrary function $f(\boldsymbol{\theta})$. The extension of the proof of the *coincidence of averages* to infinite trigonometric series is given by Arnold (1982, sec. 11).

The equality of time and spatial averages is a characteristic of *ergodic motion*. It implies the *uniform partition* of trajectories, which means that, as well as penetrating any finite region D of the torus (i.e., being dense), the time they spend in such a region is proportional to its area. To prove this property we choose $f(\boldsymbol{\theta})$ to be the characteristic function for the region, equal to one in D and zero outside of D. So if $\mu(D)$ is the area of D and $\tau(D, T, \boldsymbol{\theta}_0)$ is the total time that the trajectory stays in D in the interval $(0, T)$, we obtain from (1.54)

$$\lim_{T \to \infty} \frac{\tau(D, T, \boldsymbol{\theta}_0)}{T} = \frac{\mu(D)}{\mu(\text{torus})} = \frac{\mu(D)}{4\pi^2}. \tag{1.57}$$

This result is also true, of course, for an L torus if we define μ to be the L-dimensional volume.

1.4 Linear maps on the torus

Are there linear maps on the torus that are more interesting than mere translations? In particular, can one find motion with an isolated fixed point as in the plane? Let us consider the case of the circle. The periodic condition, that lattice points transform among themselves, cannot be satisfied if the origin is a stable fixed point. So the map would have the form

$$\theta_2' = n\theta_2 \qquad (\text{mod } 2\pi). \tag{1.58}$$

Though it is of some interest, this is not a one-to-one mapping – each θ_2' is the image of two points θ_2. This property disqualifies (1.58) as a model for the Poincaré map of a dynamical system. For a two-dimensional torus the situation is more favourable. The map

$$\begin{bmatrix} \theta_1' \\ \theta_2' \end{bmatrix} = \begin{bmatrix} m_{11} & m_{12} \\ m_{21} & m_{22} \end{bmatrix} \begin{bmatrix} \theta_1 \\ \theta_2 \end{bmatrix} \qquad (\text{mod } 2\pi) \tag{1.59}$$

will be periodic if all the elements m_{ij} of the matrix $\mathcal{M}$ are integers. The condition that the mapping be one-to-one is just that $\det \mathcal{M} = 1$.

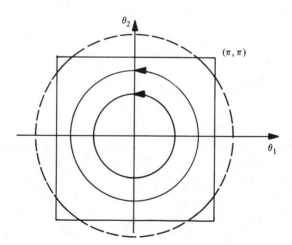

Fig. 1.8. The invariant curves for the linear map on the torus specified by (1.59) and (1.60) are circles in the $\boldsymbol{\theta}$ coordinates.

Example

$$\mathcal{M} = \begin{bmatrix} 0 & -1 \\ 1 & 0 \end{bmatrix}. \tag{1.60}$$

The eigenvalues of $\mathcal{M}$ are $\exp(\pm i\pi/2)$. The origin is a fixed point about which the plane rotates 90° for each iteration. The invariant curves for this map are circles (fig. 1.8), even when these cross the boundary of the unit cell centred on the origin.

Example. The matrix

$$\mathcal{M} = \begin{bmatrix} 2 & 1 \\ 1 & 1 \end{bmatrix} \tag{1.61}$$

generates the *Anosov automorphism*, colloquially known as *Arnold's cat map* (fig. 1.9). The origin is an unstable fixed point with real eigenvalues $\gamma_1 > 1 > \gamma_2$ equal to $(3 \pm \sqrt{5})/2$. The eigenvectors ξ_1 and ξ_2 shown in fig. 1.10 have irrationally related components since $\mathcal{M}$ is an integer matrix. The invariant helicoidal lines $t\xi_j$ are therefore dense on the torus. The points of intersection of these two lines (fig. 1.11) are called *homoclinic points*. Images of intersections are also intersections of the invariant lines, so there are an infinite number of homoclinic points. The homoclinic orbits are the only

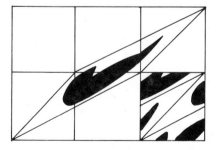

Fig. 1.9. Arnold's cat map stretches the unit square and folds it back onto itself in such a way that points initially close together become separated and vice versa.

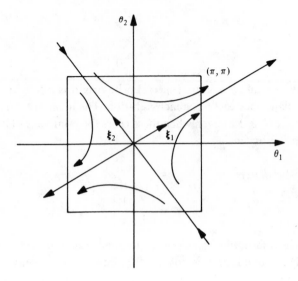

Fig. 1.10. The real eigenvectors have components that are irrationally related, so the respective orbits wind densely around the torus.

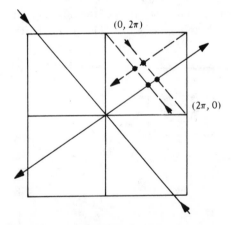

Fig. 1.11. The intersections of the incoming and outgoing separatrices are called homoclinic points. These are dense on the torus.

ones with the property that they accumulate on the origin for both forward (positive) and backward (negative) iterations of the map.

The map is linear, so any region F of the torus is stretched in the ξ_1 direction and contracted along ξ_2. After a large number n of iterations, F is deformed into a very long and thin helicoidal strip winding around the torus. For n sufficiently large, $\mathcal{M}^n F$ will intersect many times any other region G of the torus. In fact, we shall show in section 3.4 that the cat map has the property of *mixing*; that is, the ratio of the area of the intersection $G \cap \mathcal{M}^n F$ to G is the same as the ratio of the area of F to the torus itself.

The existence of isolated homoclinic points and the property of mixing are typical of the dynamics of chaotic systems. Paradoxically perhaps, they are associated with the presence of periodic orbits, the most regular form of motion! In the present example they can be trivially constructed from the observation that any point with coordinates $\boldsymbol{\theta}_0 = 2\pi(r_1/s_1, r_2/s_2)$, where r_j and s_j are mutually prime, will return to itself after a number of iterations n equal to the minimum common multiple of s_1 and s_2. Periodic orbits are the fixed points of the mapping with matrix $\mathcal{M}^n$ for some n, so each point, specified by coordinates of the above form, will have an unstable periodic orbit with eigenvalues (γ_1^n, γ_2^n). The rational numbers are dense among the reals; therefore, these periodic orbits are dense on the torus.

2
Nonlinear systems

Little can be said that applies generally to nonlinear systems. We have access mainly to local results, such as the basic existence and uniqueness theorem, which establishes the possibility of rectifying vector fields. The conditions for the theorem exclude the neighbourhoods of equilibria, exactly the regions approximated by linear fields in chapter 1. This omission is filled in by the Hartman–Grobman theorem, which guarantees the existence of a coordinate transformation that linearizes a neighbourhood of generic points of equilibrium.

There follows a discussion of Poincaré sections. In principle they reduce the study of motion near a periodic orbit to the iterations of a map in the neighbourhood of a fixed point. The linear approximation near a fixed point was shown in chapter 1 to be equivalent to a linear flow. The nontrivial extension of the Hartman–Grobman theorem to maps thus provides information on the motion surrounding an important class of periodic orbits.

Poincaré sections reduce Hamiltonian systems to a map in a phase space with two fewer dimensions. That the map preserves symplectic area (just the area, in the case of the plane) is a consequence of the Poincaré–Cartan theorem, here derived from the variational principle. We will discuss the reduction of two dimensions for a Hamiltonian system, by making it periodic in time. In particular we will see how specific choices of coordinates may preserve some of the symmetry properties of the full system.

In conclusion we will discuss two global theorems on the structural stability for motion in two dimensions. Peixoto's theorem classifies all generic dynamical systems in the plane, whereas Anosov's theorem guarantees the structural stability of the cat map.

2.1 Rectification of vector fields

It is useful to refine our mathematical terminology before discussing the general theorems presented in the following sections. We start by defining a *diffeomorphism* as a one-to-one map $\mathbf{y} = \mathbf{G}(\mathbf{x})$, where both the functions $\mathbf{G}$ and $\mathbf{G}^{-1}$ are differentiable.

A nonlinear change of coordinates can be linearized locally. There results a linear map between the vectors that act at a given point and those that act on its image. In simple physical situations, no confusion arises if we consider the points, which suffer the nonlinear transformation, and the vectors to inhabit the same Euclidean space. The discussion becomes clearer, nonetheless, if we attribute to each point $\mathbf{x}$ a complete N-dimensional vector space, containing all the vectors that may possibly act on $\mathbf{x}$. This is called the *tangent space*. Thus, a differentiable map $\mathbf{G}$ from an N-dimensional domain U to an M-dimensional range V generates a linear map from the tangent space at $\mathbf{x}$ to the tangent space at $\mathbf{y} = \mathbf{G}(\mathbf{x})$. This *tangent map DG* is defined by the condition that, given any curve $\mathbf{x}(t)$,

$$DG\left(\frac{d\mathbf{x}}{dt}\bigg|_{t=0}\right) = \frac{d}{dt}\bigg|_{t=0} (\mathbf{G}(\mathbf{x}(t))), \tag{2.1}$$

where $(d\mathbf{x}/dt)_{t=0}$ is a vector acting on the point $\mathbf{x}(0)$, while the vector on the right side acts on $\mathbf{y}(0)$. This definition makes no use of any particular choice of coordinates for $\mathbf{x}$. However, given a set of basis vectors $\{\mathbf{e}_1, \ldots, \mathbf{e}_N\}$ such that $\mathbf{x}$ has coordinates $\{x_1, \ldots, x_N\}$, the chain rule

$$\dot{y}_j = \sum_{i=1}^{N} \frac{\partial y_j}{\partial x_i} \dot{x}_i \tag{2.2}$$

identifies the matrix representation of DG with the Jacobian matrix of the nonlinear map $\mathbf{G}$.

We can now state the basic theorems, discussed and proved in Arnold (1973). There exists a diffeomorphism $\mathbf{G}:(\mathbf{x}, t) \rightarrow (\mathbf{y}, t)$, defined in some neighbourhood V of a point $(\mathbf{x}_0, t)$ of the Euclidean $(N + 1)$-dimensional space, that takes the system

$$\dot{\mathbf{x}} = \mathbf{f}(\mathbf{x}, t) \tag{2.3}$$

into the system

$$\dot{\mathbf{y}} = 0 \tag{2.4}$$

in the image $W = \mathbf{G}(V)$. The theorem is represented figuratively in fig. 2.1. Note that the diffeomorphism is defined in a space that includes t, but it does not alter the time. Thus, if $\mathbf{x}(t)$ is a solution of (2.3), $\mathbf{G}(\mathbf{x}(t))$ will be a solution of (2.4).

This *rectification theorem* implies the *existence and uniqueness theorem*, since the diffeomorphism $\mathbf{G}$ has a unique inverse. The advantage here is that we also obtain qualitatively the local structure of the orbits in the

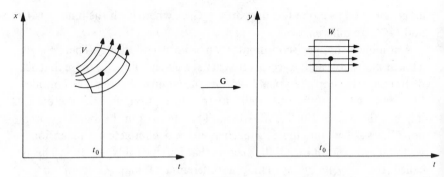

Fig. 2.1. There exists an invertible differentiable map (diffeomorphism) that transforms the trajectories in a neighbourhood V into straight lines.

neighbourhood V. To derive the corollary of the rectification theorem, which applies directly to autonomous systems, we label the phase space coordinates $\mathbf{x} = (x_1, \mathbf{X})$ and the vector field $\mathbf{f} = (f_1, \mathbf{F})$. Dividing the $N - 1$ remaining equations of the system by the first equation, we obtain the (reduced, nonautonomous) system

$$\frac{d\mathbf{X}}{dx_1} = \frac{\mathbf{F}(\mathbf{X}; x_1)}{f_1(\mathbf{X}; x_1)} \tag{2.5}$$

in the 'time' x_1, if $f_1 \neq 0$. This system can be rectified into

$$d\mathbf{Y}/dy_1 = 0, \tag{2.6}$$

where $y_1 = x_1$. A change of scale will bring $f_1 = 1$, so we obtain the full vector field as

$$\dot{\mathbf{y}} = \mathbf{e}_1 = (1, 0, \ldots, 0). \tag{2.7}$$

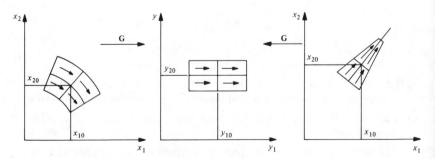

Fig. 2.2. The orbits of both the harmonic oscillator and the inverted pendulum can be rectified in a neighbourhood that excludes the origin.

If $f_1 = 0$ we can use another 'time variable' x_j, unless all the field components are zero. In conclusion, any vector field can be rectified, as long as $\mathbf{f} \neq 0$. Figure 2.2 displays the rectification of the harmonic oscillator and the inverted pendulum (after Arnold, 1973).

Exercise

Sketch the level curves of the coordinates that rectify the following vector fields:

$$\text{(a) } \mathbf{f} = x_1 \mathbf{e}_1 + 2x_2 \mathbf{e}_2,$$
$$\text{(b) } \mathbf{f} = \mathbf{e}_1 + \sin x_1 \mathbf{e}_2,$$
$$\text{(c) } \mathbf{f} = x_1 \mathbf{e}_1 + (1 - x_1^2) \mathbf{e}_2.$$

We have noted that the point where $\mathbf{f} = 0$ defines an equilibrium of the system; it is also known as a *singularity* of the vector field.

Exercises

1. Prove that vector field singularities are invariant under diffeomorphisms; that is, if $\mathbf{x}_e$ is a singular point of $\mathbf{f}(\mathbf{x})$, then $\mathbf{G}(\mathbf{x}_e)$ is a singular point of $D\mathbf{G}(\mathbf{f})$.
2. The dynamical matrix near an equilibrium is $d\mathbf{f}/d\mathbf{x}$. Show that its eigenvalues are invariant under diffeomorphisms.

2.2 The Hartman–Grobman theorem

The local structure of the orbits near singularities of the vector field is not elucidated by the rectification theorem. It is natural therefore to try to classify the types of motion around equilibria, according to the different kinds of linearized motion studied in the first chapter. The difficulty is that these depend only on whether the eigenvalue is real or complex or degenerate. On the other hand, diffeomorphisms cannot reduce related linearized motions to a standard form, as a consequence of the last exercise in section 2.1. For example, no diffeomorphism can transform the system $\dot{x} = 2x$ into $\dot{x} = x$. Thus, the partition of diffeomorphic vector fields is continuous rather than discrete.

So as not to distinguish between fields such as x and $2x$, we need a cruder notion of equivalence between vector fields. This is based on *homeomorphisms*, that is, continuous one-to-one maps with a continuous inverse, but not necessarily differentiable.

Exercise

Show that the homeomorphism $y = x|x|$ takes the system $\dot{y} = 2y$ into $\dot{x} = x$, for $x \neq 0$.

Where the homeomorphism is not differentiable, no new vector field is implicitly defined. We thus denote two systems as *topologically equivalent*, if there exists a homeomorphism $\mathbf{y} = \mathbf{G}(\mathbf{x})$ that identifies the respective flows

$$\mathbf{G}(\mathbf{x}(\mathbf{x}_0, t)) = \mathbf{y}(\mathbf{G}(\mathbf{x}_0), t) \qquad (2.8)$$

for all $\mathbf{x}_0$.

We can now enunciate the basic theorem of Hartman and Grobman: If the linear part of the field near a point of equilibrium has no zero or purely imaginary eigenvalues, then the system and its linear part are topologically equivalent within some neighbourhood of the equilibrium.

This is just as one would hope: The different kinds of linear motion studied in chapter 1 correspond to the flows near general points of equilibrium. Equilibria with no purely imaginary eigenvalues are said to be *hyperbolic*. The exception for centres with imaginary eigenvalues is easy to understand. The motion in the eigenplane was in this case seen to lie on closed ellipses, the boundary case between tending to the origin or away from it. Higher terms in the Taylor expansion of the vector field may still push it either way. However, purely imaginary eigenvalues are generic in the class of Hamiltonian systems, so that in their case the Hartman–Grobman theorem is only a partial result. A full understanding of the motion near a stable Hamiltonian system will become possible only after the discussion of the theorem of Kolmogorov, Arnold and Moser in chapter 6.

The Hartman–Grobman theorem guarantees that for every real eigenvector (or eigenplane) of the linearized system there corresponds a curve (or a two-dimensional surface) in the nonlinear system. Grouping together all the eigenvectors with a negative real part, the resulting invariant plane E^- with dimension N^- contains all the points that tend to the equilibrium $\mathbf{x}_e$ in the linearized system as $t \to \infty$. Similarly, the invariant plane E^+ with dimension N^+ contains all the $\mathbf{x} \to \mathbf{x}_e$ as $t \to -\infty$. If $\mathbf{x}_e$ is hyperbolic, then $N^- + N^+ = N$. The Hartman–Grobman theorem identifies E^- and E^+, respectively, with W^-, *the stable manifold*, and W^+, *the unstable manifold*, of the nonlinear system, while preserving the property that the points in $W^\pm$ tend to $\mathbf{x}_e$ when $t \to \mp \infty$. Though the Hartman–Grobman theorem is only local, we can obviously extend the stable manifold to include the orbits of all of its points, defined in the neighbourhood of $\mathbf{x}_e$ for $t \to -\infty$, and

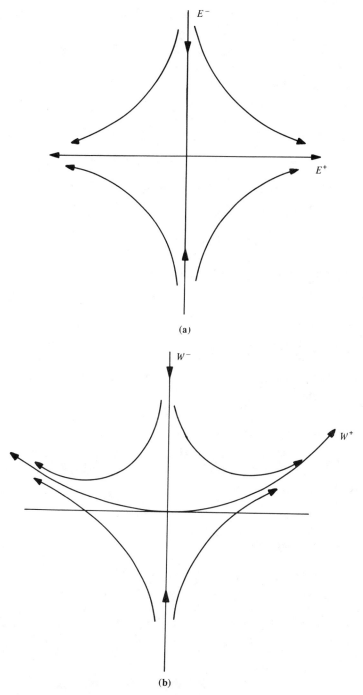

Fig. 2.3. (a) Linearized motion for the system $\dot{x} = x$, $\dot{y} = y + x^2$. (b) Stable and unstable manifolds $W^{\pm}$ for the full nonlinear system.

likewise for W^+. The *stable manifold theorem* establishes the plausible result that $W^\pm$ are tangent at $\mathbf{x}_e$ to the invariant planes $E^\pm$ and also that $W^\pm$ are differentiable as many times as the field $\mathbf{f}$ itself.

The foregoing abstract discussion becomes more tangible with a simple example and some exercises taken from Guckenheimer and Holmes (1983). Consider the system

$$\dot{x} = x, \qquad \dot{y} = -y + x^2.$$

In this case E^+ corresponds to the x axis and E^- to the y axis. Eliminating the time from the system, we obtain the differential equation

$$\frac{dy}{dx} = -\frac{y}{x} + x, \qquad \text{or} \qquad x\frac{dy}{dx} + y = x^2.$$

This can be immediately integrated, whereby

$$y = \frac{x^2}{3} + \frac{c}{x}$$

is the equation of the orbits. From the stable manifold theorem we know that W^+ is the orbit tangent to the x axis at the origin; it has $c = 0$. The stable manifold W^- is still the y axis. The graphs for the linear and the nonlinear systems are shown in figs. 2.3a and b, respectively.

Exercises

Find the points of equilibrium of the following systems and classify the respective linear parts. Start by rewriting the second-order equations as first-order systems:

$$(a) \quad \ddot{x} + \varepsilon\dot{x} - x + x^3 = 0,$$

$$(b) \quad \ddot{x} + \sin x = 0,$$

$$(c) \quad \ddot{x} + \varepsilon\dot{x}^2 + \sin x = 0,$$

$$(d) \quad \dot{x} = -x + x^2, \qquad \dot{y} = x + y,$$

$$(e) \quad \ddot{x} + \varepsilon(x^2 - 1)\dot{x} + x = 0.$$

(Where ε appears let $\varepsilon < 0$, $\varepsilon = 0$ and $\varepsilon > 0$.) Try to surmise the global structure of the stable and unstable manifolds in all these cases.

2.3 General Poincaré sections: maps

A special kind of Poincaré section has already been presented in section 1.3. In general we cut the N-dimensional phase space by an $(N-1)$-dimensional plane Σ [or more generally by an $(N-1)$-dimensional manifold]. The first return of the orbit of each point in Σ, with the same orientation with which it started off, defines a diffeomorphism $\mathbf{T}: \Sigma \to \Sigma$, known as the Poincaré map. If τ is the time that an orbit starting at $\mathbf{x}_0$ takes to return to Σ, we have

$$\mathbf{T}(\mathbf{x}_0) = \mathbf{x}(\mathbf{x}_0, \tau). \tag{2.9}$$

If Σ is a coordinate plane, (2.9) reduces to a set of $N-1$ equations.

A periodic orbit of a dynamical system generates *periodic points* of the Poincaré map; that is, for all of the intersections $\mathbf{x}_j$ of the orbit with Σ, there is a number n such that

$$\mathbf{T}^n(\mathbf{x}_j) = \mathbf{x}_j. \tag{2.10}$$

In other words, a periodic point is a *fixed point* of some power of the map. The diffeomorphism $\mathbf{T}^n$ can always be approximated by its linear part $D\mathbf{T}^n$. The *stability matrix* $\mathcal{M}$ associated with this map can then be identified with the flow matrix $\exp(\mathcal{A}t)$ for $t = 1$ of the linear system, whose dynamical matrix is $\mathcal{A} = \ln \mathcal{M}$. Each eigenvalue γ of $\mathcal{M}$ corresponds to an eigenvalue $\lambda = \ln \gamma$ of $\mathcal{A}$ and both matrices have the same eigenvectors. We can thus define stable and unstable invariant planes $E^{\pm}$ for the map $\mathbf{T}^n$ in exactly the same way as for a continuous flow. The point $\mathbf{x}_j$ will be hyperbolic if the dimensions $N^{\pm}$ of $E^{\pm}$ add up to N, that is, if $D\mathbf{T}^n$ has no eigenvalues with unit modulus.

The Hartman–Grobman theorem can be extended to maps. First we define *topological equivalence* between two maps (diffeomorphisms $\mathbf{F}$ and $\mathbf{G}$) to mean that there exists a homeomorphism $\mathbf{y} = \mathbf{H}(\mathbf{x})$ such that $\mathbf{H}(\mathbf{F}(\mathbf{x})) = \mathbf{G}(\mathbf{y})$. The theorem is then as follows: Let $\mathbf{G}$ be a diffeomorphism with a hyperbolic fixed point $\mathbf{x}_e$. Then there exists a neighbourhood of $\mathbf{x}_e$ where $\mathbf{G}$ is topologically equivalent to $D\mathbf{G}$, the linear part of $\mathbf{G}$.

The fundamental importance of this extension of the Hartman–Grobman theorem lies in the fact that periodic orbits are much more common than equilibria in dynamical systems. In fact, we have shown that they are dense in the case of the cat map.

The discrete form of the Hartman–Grobman theorem is proved in Arnold (1982), and the theorem for flows is an immediate consequence. The stable manifold theorem can also be extended to diffeomorphisms; that is, the stable and unstable manifolds $W^{\pm}$, corresponding to $E^{\pm}$, are tangent to the planes $E^{\pm}$ at $\mathbf{x}_e$; the $W^{\pm}$ surfaces are differentiable as many times as the map $\mathbf{G}$. It should be noted that the orbits of a map will be a sequence of discrete points, even if they lie on an invariant curve. The local structure of the orbits for a map on the plane with real eigenvalues is shown in fig. 2.4.

Returning now to the case of a periodic point, we note that the linearized

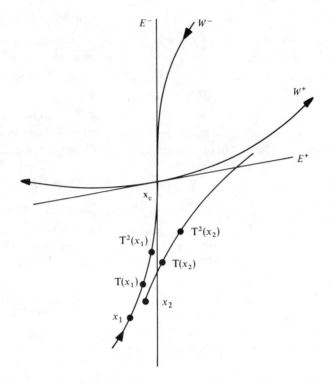

Fig. 2.4. Nonlinear maps can have stable and unstable manifolds just as continuous systems do, even though the orbits are discrete points.

transformation is simply given by the chain rule as

$$DT^n(\mathbf{x}_0) = DT(T^{n-1}(\mathbf{x}_0))DT(T^{n-2}(\mathbf{x}_0))\cdots DT(\mathbf{x}_0). \qquad (2.11)$$

In the case of the Poincaré map, however, we are still bound by the need to follow the flow, so as to work out the map. Poincaré sections are a good way to present computations of the orbits of higher-dimensional systems, but only very rarely can we directly deduce the map from the flow, as in the example at the end of this section.

The great value of Poincaré sections manifests itself in the study of generic properties of dynamical systems. For instance, we see that in the neighbourhood of a *hyperbolic periodic orbit* γ, corresponding to a hyperbolic periodic point of a Poincaré section, there will be stable and unstable manifolds $W(\gamma)$, such that all the orbits on $W^-(\gamma)$ approach γ as $t \to \infty$, whereas all orbits in $W^+(\gamma)$ approach γ as $t \to -\infty$. The relation

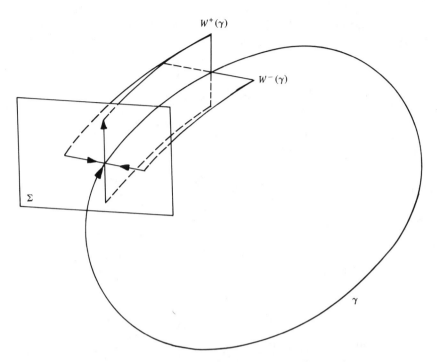

Fig. 2.5. The stable and unstable manifolds of a Poincaré map are the sections of two-dimensional surfaces in phase space that intersect along the periodic orbit.

between the full stable and unstable manifolds and those on the Poincaré section is sketched in fig. 2.5.

We conclude this section by identifying *stroboscopic maps*, derived from the observation of the flow at regular intervals, with the Poincaré map. We have already worked with stroboscopic maps in the case of autonomous linear systems, in which case they are isomorphic to their parent flows. *Forced oscillations*, where the field is periodic in time, that is, $\mathbf{f}(t + \tau) = \mathbf{f}(t)$ for all t, lead to a much richer relationship between map and flow. We can always consider the iterations of the map resulting from the flow for each period τ as the result of successive sections at the planes $\theta_0 + 2m\pi$ for the autonomous system

$$\dot{\mathbf{x}} = \mathbf{f}(\mathbf{x}, \theta\tau/2\pi), \qquad \dot{\theta} = 2\pi/\tau. \tag{2.12}$$

The following example of a forced linear oscillator

$$\ddot{x} + 2\beta\dot{x} + x = \gamma \cos \omega t$$

at resonance, $\omega = |1 - \beta^2|^{1/2}$, is borrowed from Guckenheimer and Holmes (1983). It can be treated as the system

$$\begin{bmatrix} \dot{x}_1 \\ \dot{x}_2 \end{bmatrix} = \begin{bmatrix} 0 & 1 \\ -1 & -2\beta \end{bmatrix} \begin{bmatrix} x_1 \\ x_2 \end{bmatrix} + \begin{bmatrix} 0 \\ \gamma \cos \theta \end{bmatrix},$$

$$\dot{\theta} = \omega.$$

The general solution of the second-order equation is known to have the form

$$x(t) = \exp(-\beta t)(C_1 \cos \omega t + C_2 \sin \omega t) + A \cos \omega t + B \sin \omega t,$$

where the constants C_1 and C_2 are determined by the initial conditions as

$$C_1 = x_0 - A, \qquad C_2 = -B + [\dot{x}_0 + \beta(x_0 - A)]/\omega.$$

Identifying $x_1(t) = x(t)$ and $x_2(t) = \dot{x}(t)$, we obtain the Poincaré map $\mathbf{x}(\mathbf{x}_0, t = 2\pi/\omega)$ as

$$x_1 = A + (x_{10} - A)\exp(-2\pi\beta/\omega),$$

$$x_2 = \omega B + (x_{20} - \omega B)\exp(-2\pi\beta/\omega).$$

This is a linear map with degenerate eigenvalues $\exp(-2\pi\beta/\omega)$. Therefore, the map orbits approach the fixed point $(A, \omega B)$ radially (fig. 2.6). The flow

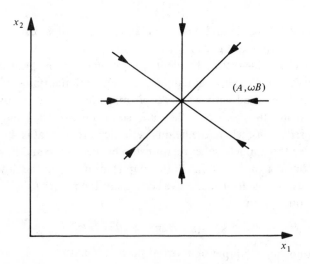

Fig. 2.6. The fixed point of the Poincaré map for the system obtained from the equation $\ddot{x} + 2\beta\dot{x} + x = \gamma \cos \omega t$ is not at the origin. The map orbits approach this point radially.

tends asymptotically to a periodic orbit. Note that even in this simple case the map depends on the initial time or phase θ_0, taken here to be zero.

2.4 Poincaré sections of Hamiltonian systems

The conservation of energy in a Hamiltonian system enables us to reduce the phase space of a Poincaré section by two dimensions rather than one. This follows from the restriction of the Poincare section (2.9) to *the energy surface* or *energy shell*,

$$H(\mathbf{p}, \mathbf{q}) = E. \tag{2.13}$$

Thus, if Σ is the plane $q_1 = 0$, we can use $(p_2, \ldots, p_L, q_2, \ldots, q_L) = (\mathbf{P}, \mathbf{Q})$ as coordinates for the intersection of Σ with the energy surface, and determine p_1 from (2.13).

A fundamental property of Hamiltonian Poincaré maps is that they are canonical transformations, in the same way as are the stroboscopic maps generated by Hamiltonian flows. Invariance of the symplectic area for differential parallelograms implies the conservation of the symplectic area S for any closed loop in the phase space

$$S = \sum_{l=1}^{L} S_l, \tag{2.14}$$

where each S_l is the area of the projection of the loop onto the lth

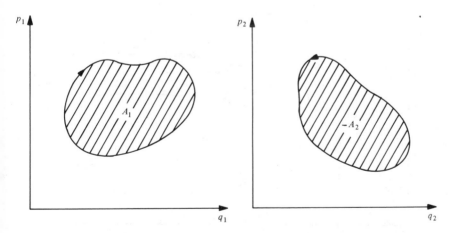

Fig. 2.7. The symplectic area for a closed circuit is the algebraic sum of the areas of its projections onto all the canonical conjugate planes. It is invariant with respect to Hamiltonian flows.

conjugated plane,

$$S_l = \oint p_l \, dq_l, \tag{2.15}$$

as shown in fig. 2.7. (Plane Poincaré maps are area preserving.) The symplectic area S is also known as the *reduced action*. Its invariance is a consequence of the *Poincaré–Cartan integral theorem*: Given a tube of trajectories in extended phase $(\mathbf{p}, \mathbf{q}, t)$ generated by a closed curve Γ_1, then any other irreducible curve Γ_2 around the same tube will have the same *full action*:

$$\oint_{\Gamma_1} [\mathbf{p} \cdot d\mathbf{q} - H \, dt] = \oint_{\Gamma_2} [\mathbf{p} \cdot d\mathbf{q} - H \, dt]. \tag{2.16}$$

Figure 2.8 shows a tube of trajectories for a system with a single freedom. It is important to note that the theorem in no way restricts each curve Γ_j to have a definite time or a given energy E. The imposition of either of these conditions leads immediately to the conservation of the reduced action

$$\oint_{\Gamma_1} \mathbf{p} \cdot d\mathbf{q} = \oint_{\Gamma_2} \mathbf{p} \cdot d\mathbf{q}. \tag{2.17}$$

Any closed circuit Γ_1 in a Poincaré section Σ defines a tube of trajectories that intersects Σ again in a new circuit Γ_2. Since all these orbits have the same energy by construction, (2.17) holds.

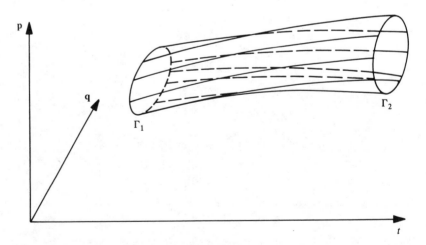

Fig. 2.8. The trajectories that start out from a closed curve Γ_1 define a tube of trajectories; Γ_2 is any other curve obtained by cutting this tube.

The Poincaré–Cartan integral theorem is one of the cornerstones of classical mechanics. Arnold (1978) uses it to derive Hamilton's *variational principle*. Yet the latter is the more familiar starting point in physics courses, so it is worthwhile outlining the deduction in the opposite direction. The *variational principle* states that, given any one-parameter family of curves γ_λ joining the lines (q_1, t_1) to (q_2, t_2) in extended phase space such that γ_0 is a trajectory of the system, then

$$\int_{\gamma_\lambda} [\mathbf{p} \cdot d\mathbf{q} - H \, dt] - \int_{\gamma_0} [\mathbf{p} \cdot d\mathbf{q} - H \, dt] = O(\lambda^2). \tag{2.18}$$

Here we introduce the standard notation, that a function $f(x) = O(x)$ if

$$\lim_{x \to 0} \left| \frac{f(x)}{x} \right| < \infty. \tag{2.19}$$

Let us now define a one-parameter family of trajectories $\gamma'(\lambda)$, so that $\gamma'(0) = \gamma_0$, and let Γ be an arbitrary curve joining $(\mathbf{q}_1, t)$ and $(\mathbf{q}_2, t)$, embedded in the resulting two-dimensional surface. This construction is shown in fig. 2.9 in the case of one freedom. If Λ is the largest value taken by λ along Γ, we can calculate the closed integral

$$\sigma \equiv \int_{\gamma_0} [\mathbf{p} \cdot d\mathbf{q} - H \, dt] - \int_{\Gamma} [\mathbf{p} \cdot d\mathbf{q} - H \, dt] \tag{2.20}$$

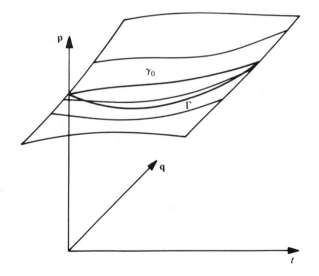

Fig. 2.9. In this construction γ_0 is a trajectory and Γ is some curve contained in the surface $\gamma'(\lambda)$, with the same end points as γ_0.

by subdividing it into the n closed integrals

$$\sigma = \sum_{j=1}^{n} \oint_j [\mathbf{p} \cdot d\mathbf{q} - H\, dt], \qquad (2.21)$$

obtained by starting out from Γ along $\gamma'(j\Lambda/n)$ until meeting Λ again and then returning along $\gamma'((j+1)\Lambda/n)$. Each integral in (2.21) can be interpreted as the difference in total action between a trajectory and a nearby path. Thus, the variational principle (2.18) implies that

$$\sigma = \lim_{n \to \infty} \sum_{j=1}^{n} O\left(\left(\frac{\Lambda}{n}\right)^2\right) = 0. \qquad (2.22)$$

Consider now an arbitrary *reducible curve* inscribed in a one-parameter family of trajectories, which means that the curve can be continuously shrunk to a point, without leaving the two-dimensional surface. The total action for such a curve will be zero as a consequence of (2.22). This is not the case of the two curves Γ_1 and Γ_2 specified in the Poincaré–Cartan theorem, but we can construct a reducible curve by cutting the tube of trajectories in fig. 2.8 and joining Γ_1 and Γ_2 by the two paths along the cut. The integrals of the action along both paths on either side of the cut cancel out, leaving

$$\sigma = \sigma_{\Gamma_1} - \sigma_{\Gamma_2} = 0. \qquad (2.23)$$

We conclude this section with a discussion of the local motion in the neighbourhood of periodic orbits of Hamiltonian systems. A Poincaré section reduces the flow to a canonical map with two fewer dimensions. If the map is linearized, it can be identified with the linear flow derived from a quadratic Hamiltonian belonging to one of the categories classified in section 1.2. If the original system has two freedoms, the reduced system has only one; then the generic possibilities are real or purely imaginary eigenvalues (classes 1 and 2 in section 1.2). The Hartman–Grobman theorem guarantees that in the former case the nonlinear map can be transformed to its linear part by a homeomorphism. For three or more freedoms, it is also possible to have a reduced quadratic Hamiltonian in one of the classes 3 and 5, involving, respectively, complex eigenvalues or real degenerate eigenvalues. In these cases the corresponding nonlinear maps are topologically equivalent to their linear part.

2.5 Reduction to time-periodic systems

We found in section 2.3 that the stroboscopic map of a time-periodic system can be identified with the Poincaré map of an autonomous system augmented by one dimension. We shall now investigate the converse procedure, namely, the reduction of autonomous systems to time-dependent systems of fewer dimensions, in the vicinity of a periodic orbit.

For general systems this task is trivial in principle. All we have to do is change the coordinate system $\mathbf{x} \rightarrow (\theta, \mathbf{X})$ with vector field $(f_\theta, \mathbf{F})$ in such a way that the closed orbit (of period τ) becomes the θ coordinate line $(\theta, \mathbf{0})$ with velocity

$$\dot{\theta} = \omega = 2\pi/\tau. \tag{2.24}$$

Neighbouring orbits will not generally coincide with the closed lines $(\theta, \mathbf{X}_0)$. The construction for a two-dimensional system showing the periodic orbit and one of its neighbours is drawn in fig. 2.10. Because of continuity, there will be a region surrounding the periodic orbit where

$$\dot{\theta} = f_\theta(\theta, \mathbf{X}) \neq 0. \tag{2.25}$$

In this neighbourhood we can define the reduced system

$$\frac{d\mathbf{X}}{d\theta} = \frac{\mathbf{F}(\mathbf{X}; \theta)}{f_\theta(\mathbf{X}; \theta)}, \tag{2.26}$$

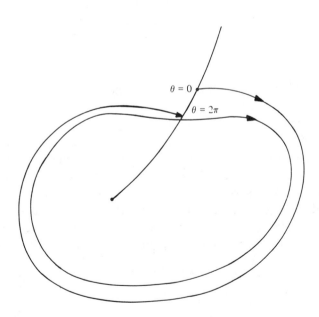

Fig. 2.10. Orbits near a periodic orbit are not generally closed lines, but in this region the angular coordinate satisfies $\dot{\theta} \neq 0$.

(2π)-periodic in the 'time' θ. The strosboscopic map $\theta = 2m\pi$ of the reduced system coincides with the Poincaré map of the full system (fig. 2.10).

A Hamiltonian system requires more careful treatment if we are to ensure that the reduced system is also Hamiltonian. Let us for the moment suppose that canonical coordinates $(J, \mathbf{P}, \theta, \mathbf{Q})$ can be found, where θ has the same properties as in the preceding paragraph. Then condition (2.25), that is,

$$\dot{\theta} = \partial H / \partial J \neq 0 \tag{2.27}$$

near the orbit, implies that we can invert the relation $H(\mathbf{p}, \mathbf{q}) = E$ to obtain the well-behaved function

$$J = K(\mathbf{P}, \mathbf{Q}; \theta; E). \tag{2.28}$$

Consider now the variational system. It identifies trajectories as the paths that minimize the complete action among all possible paths. In particular, these include all the paths with the same energy as both end-points of the trajectory. But for these paths the full action equals the reduced action:

$$\int [\mathbf{p} \cdot d\mathbf{q} - H \, dt] = \int \mathbf{p} \cdot d\mathbf{q} = \int [\mathbf{P} \cdot d\mathbf{Q} - K d(-\theta)]. \tag{2.29}$$

The isomorphism between the first and the last term in (2.29) then guarantees that the orbits in the energy shell satisfy Hamilton's equations

$$\frac{d\mathbf{P}}{dt} = -\frac{\partial K}{\partial \mathbf{Q}}, \qquad \frac{d\mathbf{Q}}{dt} = \frac{\partial K}{\partial \mathbf{P}}, \tag{2.30}$$

with $t = -\theta$. Thus, an autonomous Hamiltonian system can be reduced by a whole freedom near a periodic orbit. The stroboscopic map for the reduced system can be identified with the Poincaré map for the full system.

Reduction of the Hamiltonian was achieved under the assumption that suitable canonical coordinates could be found, which must now be verified. The first point to note is that periodic orbits generally appear in Hamiltonian systems within one-parameter families. This is easily shown by considering the one-parameter family of Poincaré sections, obtained by varying the energy. The intersection of a given periodic orbit of energy E with its Poincaré section is a fixed point of the map:

$$\mathbf{T}^n(\mathbf{X}, E) = \mathbf{X}. \tag{2.31}$$

Differentiating (2.31) with respect to E, we obtain

$$\frac{d\mathbf{X}}{dE} = (\mathbf{1} - \mathcal{M}^n)^{-1} \frac{d\mathbf{T}^n}{dE}, \tag{2.32}$$

where $\mathscr{M}'' = D\mathbf{T}''$. So we can use the implicit function theorem to define the family of fixed points $\mathbf{X}(E)$, except at the *resonant energies* where

$$\det(\mathscr{M}'' - \mathbf{1}) = 0. \tag{2.33}$$

We shall see in chapter 4 that at these energies the periodic orbit family *bifurcates*; that is, there is a coalescence of orbit families. In some cases (integrable systems) the periodic orbits come in families with L or more parameters, but in the following discussion we shall consider only the two-dimensional annulus-γ obtained by varying a single energy-dependent parameter.

The ideal coordinate system, to study the neighbourhood of periodic orbits, embeds the annulus-γ in a coordinate surface. To find what kind of surface this is, we check the skew product of two independent vectors tangent to γ. Choosing $\eta = (-\partial H/\partial \mathbf{q}, \partial H/\partial \mathbf{p})$, the velocity vector tangent to the periodic orbit, and $\xi = (d\mathbf{p}, d\mathbf{q})$, we have

$$\eta \wedge \xi = \frac{\partial H}{\partial \mathbf{p}} \cdot d\mathbf{p} + \frac{\partial H}{\partial \mathbf{q}} \cdot d\mathbf{q} = dH(\xi) \neq 0 \tag{2.34}$$

by construction. According to conditions (1.46) we can then choose ξ and η to be in the new (p_1, q_1) plane.

By contrast, any vector ξ' chosen in the tangent plane of the energy surface along the periodic orbit has $dH(\xi') = 0$; therefore, its skew product with η is zero. The result of the exercise on page 13 is that any vector ξ defines a unique $(2L - 1)$-dimensional plane with the property that all the vectors in it have a null skew product with ξ (its skew-orthogonal plane). Hence, for any choice of ξ in γ, as in (2.34), the intersection of its skew-orthogonal plane with the tangent plane to the energy surface is Γ, a $(2L - 2)$-dimensional plane that can be taken as the basis for the remaining $(\mathbf{P}, \mathbf{Q})$ coordinates. Figure 2.11 shows a three-dimensional reduction of the resulting geometry, in which γ is fully two-dimensional, the energy surface appears as two-dimensional and Γ is reduced to a single dimension.

Taking radial and angular coordinates in the (p_1, q_1) plane, so that the periodic orbits are coordinate lines, we achieve the required canonical transformation. As a bonus we also obtain

$$\left.\frac{\partial H}{\partial \mathbf{P}}\right|_\gamma = \left.\frac{\partial H}{\partial \mathbf{Q}}\right|_\gamma = 0, \tag{2.35}$$

that is, local invariance of the Hamiltonian with respect to the transverse coordinates $(\mathbf{P}, \mathbf{Q})$. However, it is not generally possible to find a transformation that makes the Hamiltonian completely independent of the

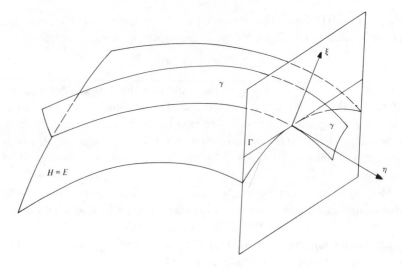

Fig. 2.11. γ is the two-dimensional annulus formed by the one-parameter family of periodic orbits. The skew-orthogonal plane Γ to the flow vector η of a periodic orbit is tangent to the energy surface $H(\mathbf{p}, \mathbf{q}) = E$.

transverse coordinates, for then all orbits would lie in two-dimensional $(\mathbf{P}, \mathbf{Q}) = $ const planes; that is, all orbits would be periodic. At best we can choose the Hamiltonian to be independent of $(\mathbf{P}, \mathbf{Q})$ for $|\theta| < \alpha < \pi$, so that the $(\theta = 0, J = \text{const})$ plane coincides with the Poincaré section.

The generality of the above reduction can distort symmetries of the original system, which may be worth preserving. There is a simpler alternative, when the projection of the closed orbit onto one of the coordinate planes does not intersect itself (fig. 2.12). We can then choose this to be the (p_1, q_1) plane and transform to (J, θ) coordinates in it, so that $2\pi J$ is the area of each projected closed orbit and $\dot{\theta} = \text{const}$ along each orbit. Then $(\mathbf{P}, \mathbf{Q})$ will be the remaining coordinates of the untransformed system.

Suppose now that the system has a special symmetry. As an example we will examine the important case in which it has *time-reversal symmetry*; namely, the motion projected onto the $\mathbf{q}$ coordinate plane has the property that changing the velocity $\dot{\mathbf{q}} \rightarrow -\dot{\mathbf{q}}$ amounts to running the system backward in time. In particular this is obtained for p-symmetric Hamiltonians:

$$H(\mathbf{p}, \mathbf{q}) = H(-\mathbf{p}, \mathbf{q}). \tag{2.36}$$

Consider now running a symmetric periodic orbit forward and backward in time from the point where $p_1 = 0$. Because of the symmetry of the

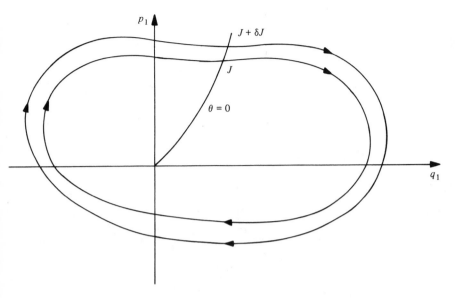

Fig. 2.12. The family of periodic orbits can be projected onto the (p_1, q_1) plane, and the variable J can be defined so that $2\pi J$ is the area of a periodic orbit.

Hamiltonian and the fact that $\dot{\theta}$ is constant, we shall have $-\theta(t) = \theta(-t)$, so $p_1(-\theta, J, -\mathbf{P}, \mathbf{Q}) = -p_1(\theta, J, \mathbf{P}, \mathbf{Q})$ and $q_1(-\theta, J, -\mathbf{P}, \mathbf{Q}) = q_1(\theta, J, \mathbf{P}, \mathbf{Q})$. Thus, the new coordinates preserve symmetry for the full Hamiltonian,

$$H(J, \theta, \mathbf{P}, \mathbf{Q}) = H(J, -\theta, -\mathbf{P}, \mathbf{Q}), \qquad (2.37)$$

and for the reduced Hamiltonian,

$$K(\mathbf{P}, \mathbf{Q}; \theta; E) = K(-\mathbf{P}, \mathbf{Q}; -\theta; E). \qquad (2.38)$$

Since the Poincaré map $\mathbf{T}$ is the stroboscopic map for the reduced flow, in which θ is taken from $0 \to -2\pi$, the inverse $\mathbf{T}^{-1}$ is the map for which θ is taken from $0 \to +2\pi$. Therefore, the inverse Poincaré map is obtained by reflecting the coordinate $\mathbf{P}$, propagating the point forward in time and reflecting $\mathbf{P}$ back again. Defining the reflection

$$\mathbf{R} : (\mathbf{P}, \mathbf{Q}) \to (-\mathbf{P}, \mathbf{Q}), \qquad (2.39)$$

we find that

$$\mathbf{T}^{-1} = \mathbf{R} \cdot \mathbf{T} \cdot \mathbf{R} \qquad (2.40)$$

as shown in fig. 2.13. This property of $\mathbf{T}$ with respect to an *inversion* $\mathbf{R}$, for

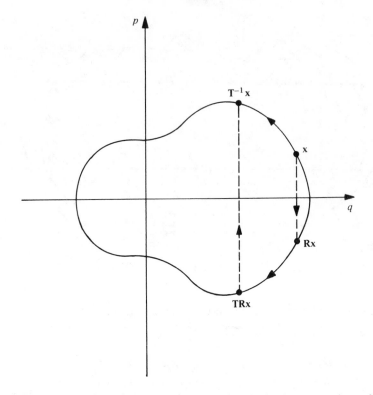

Fig. 2.13. The inverse of a reversible map can be obtained as a product of the map **T** and the symmetry **R**.

which

$$\mathbf{R} \cdot \mathbf{R} = \mathbf{1} \qquad (2.41)$$

where **1** is the identity map, defines a *reversible map* according to Greene, Mackay, Vivaldi and Feigenbaum (1981).

Exercises

1. Show that reversible maps can always be decomposed as the product of two inversions **R** and **TR**.
2. Show that the property of reversibility is invariant with respect to coordinate transformations.

2.6 Structural stability

The Hartman–Grobman theorem presented in section 2.2 establishes the topological equivalence of some neighbourhood of an equilibrium of a nonlinear system with its linear part. Nondegenerate hyperbolic linear systems with equivalent eigenvalue structures – that is, the same number of positive, negative or complex eigenvalues – can be taken into each other by homeomorphisms. It follows that all nondegenerate hyperbolic nonlinear systems with the same eigenvalue structure form a class of topologically equivalent systems. A sufficiently small perturbation of such a system will turn it into a new system with the same eigenvalue structure as it had before and hence the same equivalence class. The system is then said to be *structurally stable* in a neighbourhood of the equilibrium.

The notion of topological equivalence is too restrictive to be taken as a basis for the generalization of the definition of structural stability to the neighbourhood of periodic orbits. We have seen in the preceding section that a one-parameter family of systems will generally support one-parameter families of periodic orbits. However, there is no restriction placed on the period of these closed orbits, which must be equal in order to be topologically equivalent. Following Arnold (1982), we thus define two systems to be *topologically (orbitally) equivalent* (in some region of phase space) if there exists a homeomorphism that associates the orbits of both systems. This definition does not imply equal time intervals for the equivalent sections of each pair of orbits. Finally we define a system to be *structurally stable* if it is contained in a neighbourhood of systems topologically (orbitally) equivalent to it. This definition is inclusive of the neighbourhoods of nondegenerate hyperbolic points already treated.

The conjecture that the concept of structural stability permits a global classification of 'typical' dynamical systems is borne out for two-dimensional autonomous systems. This is the content of Peixoto's theorem (Peixoto and Peixoto, 1959): A plane dynamical system with an r-times differentiable vector field, which points in (out) of every point along a compact boundary, is structurally stable in the bounded region if and only if

1. the number of points of equilibrium and periodic orbits are finite and all are hyperbolic; and
2. there are no orbits joining a saddle point to another or to itself.

The structurally stable systems form a dense open subset in the set of systems with r-times differentiable fields and the same boundary condition.

This statement means that arbitrarily small perturbations exist which make any system structurally stable. This extremely strong result, for systems restricted only by the two conditions of Peixoto's theorem, is a consequence of the small dimensionality of the phase space. In fact, the theorem can be generalized to systems defined on compact surfaces such as

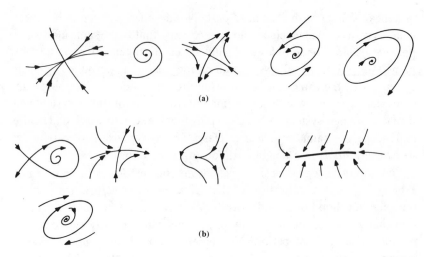

Fig. 2.14. Phase diagrams for (a) structurally stable systems in the plane and (b) structurally unstable systems, according to Peixoto's theorem.

spheres or tori (Peixoto, 1962). The orbits already have one dimension; since they never cross, there remains only one dimension in which they can get tangled up. Figure 2.14 shows examples of (a) structurally stable and (b) unstable systems in the plane.

For flows in three or more dimensions, or equivalently for maps of two or more dimensions, the situation becomes more complicated. In particular Smale constructed in 1965 an example, discussed in Arnold (1982), of a three-dimensional diffeomorphism whose neighbourhood contains no structurally stable systems.

It is worth emphasizing that we are dealing here with typical systems not generated by a Hamiltonian. For instance, we found in section 1.2 that elliptic points (centres) are structurally stable with respect to linear perturbations of the Hamiltonian, but they are unstable with respect to a general linear perturbation, or even a nonlinear Hamiltonian perturbation, as we shall see in chapter 6. The reader will find preparatory material and a sketched partial proof of Peixoto's theorem in Guckenheimer and Holmes (1983), a book dealing mainly with dissipative systems.

Exercise

Show that a Hamiltonian system in the plane never satisfies the conditions of Peixoto's theorem.

An important exception to the difficulty of achieving global understanding of the structure of two-dimensional maps is furnished by Anosov's theorem, demonstrated in Arnold (1982): The cat map is structurally stable. In other words, any diffeomorphism on the torus sufficiently close to the cat map can be taken into it by a homeomorphism.

These systems bear no resemblance to those established by Peixoto's theorem to be structurally stable – they are the prototype of chaotic conservative systems. Typically, conservative diffeomorphisms exhibit chaotic regions mixed up with regular regions, as we shall see. It is not surprising, therefore, that no statement can be made about making arbitrary maps on the torus structurally stable, in contrast to Peixoto's theorem. For example, the elliptic linear map presented in section 1.4 is neither structurally stable nor close to a linear hyperbolic map.

3

Chaotic motion

The flow in an autonomous Hamiltonian system with one freedom is even more restricted than the generic motion categorized by Peixoto's theorem. The constancy of the Hamiltonian holds all the orbits to its level curves. For a bound system most of these will be closed loops, the exception being the levels of saddle points, that is, unstable equilibria of the Hamiltonian. The corresponding orbits either start out at an unstable point as $t \to -\infty$ and return to the same point as $t \to +\infty$, or go on to another unstable point. These two cases, known respectively as *homoclinic* and *heteroclinic* orbits, are shown in figs. 3.1a and b [under the potential curves $V(q)$ that generate them, if the Hamiltonian has the form $p^2 + V(q)$]. The homoclinic orbit of the pendulum (fig. 3.1b) is known as the *separatrix* between the small *librations* of the system and the full *rotations* arising at higher energies.

Consider making a small perturbation to this system, that is, coupling to another degree of freedom or, equivalently, adding an oscillatory forcing term to the Hamiltonian. Far from the unstable equilibrium, an orbit close to the separatrix will feel mainly the unperturbed Hamiltonian. But this gives a zero force near the saddle, so the perturbation becomes dominant near the unstable point: It can switch librations into rotations and back again. Successive switches will be uncorrelated, because the periods of rotations and librations lying close to the separatrix vary infinitely. The resulting motion can be classified as chaotic, albeit constrained to a narrow strip of phase space. It is bounded by closed curves, which survive in the Poincaré map of the system, according to the theorem of Kolmogorov, Arnold and Moser to be discussed in chapter 6. Of course, the chaotic region widens as the perturbation parameter is increased.

The separatrices correspond to the stable and unstable manifolds of the equilibrium, which continue to exist in the Poincaré map. Far from the equilibrium their extensions usually cross transversely at discrete *homoclinic (heteroclinic) points*. We begin this chapter with the study of Melnikov's analytic criterion for such crossings to occur. Then we analyse the motion near homoclinic points, introducing the method of symbolic dynamics. The result is the Smale–Birkhoff theorem, which establishes the

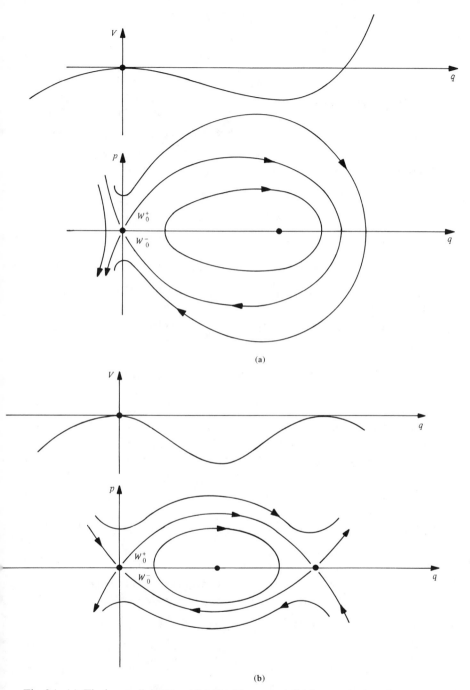

Fig. 3.1. (a) The homoclinic W_0 orbit is doubly asymptotic, for $t \to \pm \infty$, to the same unstable equilibrium. (b) A heteroclinic orbit tends asymptotically to one unstable equilibrium for $t \to -\infty$ and to another for $t \to \infty$.

existence of dense (chaotic) orbits and an infinite number of periodic orbits, such as were found in the cat map.

In the remainder of this chapter we discuss rudiments of ergodic theory. The definition of invariant measure on the phase space leads to a precise theory of what happens to 'almost all orbits'. After a preliminary study of information theory, metric entropy is introduced and connected with the local Lyapunov exponent. The last section joins many strands of the chapter in the (heuristic) uniformity principle for periodic orbits.

3.1 Melnikov's method

Consider a one-freedom system with the Hamiltonian

$$H(\mathbf{x}, t) = H_0(\mathbf{x}) + \varepsilon H_1(\mathbf{x}, t), \tag{3.1}$$

where H_1 is time periodic with period τ. We take H_0 to have an unstable point at the origin, whose stable and unstable separatrices $W_0^{\pm}$ form a smooth curve, as in fig. 3.1a. We can define a continuous set of Poincaré sections Σ_{t_0}; all the corresponding maps

$$\mathbf{x} \to \mathbf{x}'_{\varepsilon, t_0} = \mathbf{x}'_{\varepsilon}(\mathbf{x}(t_0), t_0 + \tau) \tag{3.2}$$

will have an unstable fixed point close to the origin, as was shown in section 2.5.

The objective is to understand the way that the separatrices $W_{\varepsilon, t_0}^{\pm}$ extend and in particular whether they cross transversely. The method is perturbative; that is, we use the fact that $W_{\varepsilon, t_0}^{\pm}$ will remain close to $W_0^{\pm}$ within appropriate time intervals, $[t_0, \infty)$ for the stable manifold and $(-\infty, t_0]$ for the unstable manifold. For instance, it is proved in Guckenheimer and Holmes (1983) that the distance between the orbits on the perturbed stable separatrix $\mathbf{x}_{\varepsilon}^-(\mathbf{x}_{\varepsilon}^-(t_0), t)$ and the unperturbed separatrix $\mathbf{x}_0(\mathbf{x}_0, t - t_0)$ is uniformly of order ε in (t_0, ∞).

Let us consider now a point $\mathbf{x}_0(0)$ on the unperturbed separatrix and the straight line π, defined by the gradient $\partial H_0/\partial \mathbf{x}$ at this point. The line will intersect the perturbed stable and unstable manifolds at the two points $\mathbf{x}_{\varepsilon, t_0}^{\pm}(t_0)$, as shown in fig. 3.2. We define the *Melnikov function* as

$$M(t_0) = \varepsilon^{-1}[H_0(\mathbf{x}_{\varepsilon, t_0}^+(t_0)) - H_0(\mathbf{x}_{\varepsilon, t_0}^-(t_0))]. \tag{3.3}$$

Suppose now that the separatrices of the perturbed Poincaré map, corresponding to Σ_{t_0}, do interesect transversely at some point h_{t_0} above $\mathbf{x}_0$ in fig. 3.2. Its image by the map will be another point of intersection below the line π, a homoclinic point $h_{t_0 + \tau}$ of the section $\Sigma_{t_0 + \tau}$ (fig. 3.3). Considering

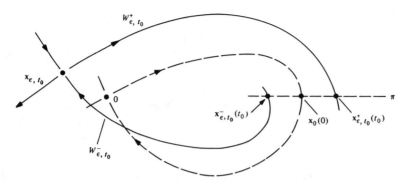

Fig. 3.2. The perturbed system also has stable and unstable manifolds $W_{\varepsilon t_0}^{\pm}$ lying close to the unperturbed separatrix (the dashed line). A straight line normal to the separatrix at $x_0(0)$ intersects both the stable and the unstable manifolds.

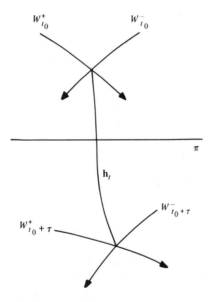

Fig. 3.3. If there exists a homoclinic point to one side of the line π, its image lies on the other side. The Poincaré sections depend on the time, and there will be a t for which the homoclinic point lies exactly on π.

all the continuum of sections between Σ_{t_0} and $\Sigma_{t_0+\tau}$, we obtain a continuous locus of points $\mathbf{h}_t$ (the homoclinic orbit of the full system) joining $\mathbf{h}_{t_0}$ and $\mathbf{h}_{t_0+\tau}$. Hence, there will be a time for which $\mathbf{h}_t$ lies on the line π, that is, for which $M(t) = 0$.

The existence of a zero of the Melnikov function amounts to the existence of a homoclinic point. Transverse intersections of the separatrices give rise to simple zeros (with $dM/dt \neq 0$). If the system is 'integrable', so that $W^{\pm}$ join smoothly, the Melnikov function is identically zero. In this case the homoclinic points are nonisolated. Usually we refer to transverse intersections of separatrices simply as homoclinic points. They are the only points along the separatrices that tend to the equilibrium both as $t \to \infty$ and as $t \to -\infty$.

We can calculate directly the $\varepsilon \to 0$ limit of the Melnikov function from H_0, H_1 and the knowledge of the unperturbed flow. All that is required is to rewrite the Melnikov function in the form

$$\varepsilon M(t_0) = \int_{-\infty}^{t} \dot{H}_0(\mathbf{x}_{\varepsilon,t_0}^{+}(t)) \, dt + \int_{t_0}^{\infty} \dot{H}_0(\mathbf{x}_{\varepsilon,t_0}^{-}(t)) \, dt, \tag{3.4}$$

which follows from the observation that $\mathbf{x}_{\varepsilon,t_0}^{+}(-\infty) = \mathbf{x}_{\varepsilon,t_0}^{-}(\infty)$, the fixed point itself. Defining the *Poisson bracket*

$$\{F, G\} = \frac{\partial F}{\partial q} \frac{\partial G}{\partial p} - \frac{\partial G}{\partial q} \frac{\partial F}{\partial p} \tag{3.5}$$

of the functions F and G and using the equation

$$\dot{H}_0 = \{H_0, H\} = \varepsilon \{H_0, H_1\} \tag{3.6}$$

(see Goldstein, 1980), we obtain

$$M(t_0) = \int_{-\infty}^{\infty} \{H_0(\mathbf{x}_0(t - t_0)), H_1(\mathbf{x}_0(t - t_0, t)\} \, dt + O(\varepsilon), \tag{3.7}$$

since the difference between $\mathbf{x}_{\varepsilon,t_0}^{\pm}(t - t_0)$ and $\mathbf{x}_0(t - t_0)$ is of $O(\varepsilon)$ within each integral in (3.4).

Thus, the calculation of the integral (3.7), for $\mathbf{x}_0$ along the unperturbed separatrix and for all t_0 in the period, will reveal whether the system has a homoclinic point. The symmetry in the integral may even be sufficient to answer the question. For example, consider $H_0 = p^2/2 + q^3/3 - q^2/2$ (fig. 3.1a) perturbed by $H_1 = (p^2/2)\cos(2\pi t/\tau)$. The Poisson bracket $\{H_0, H_1\} = (q^2 - q)p\cos(2\pi t/\tau)$, and the Melnikov function becomes

$$M(t_0) = \int_{-\infty}^{\infty} (q_0^2(t) - q_0(t))p_0(t)\cos\left(\frac{2\pi(t + t_0)}{\tau}\right) dt. \tag{3.8}$$

If we choose $p_0(0) = 0$, $q_0(0) = \frac{3}{2}$, the integrand will be antisymmetric in t for $t_0 = 0$ (fig. 3.4). Thus, $M(0) = 0$ and it is easy to verify that $M(t_0)$ is

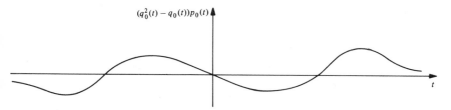

Fig. 3.4. The choice of $t_0 = 0$ makes the integrand of the Melinkov function in the example (3.8) antisymmetric, so $M(0) = 0$.

not identically zero. The perturbed system therefore has isolated homoclinic points.

The Melnikov method can be extended to deal with non-Hamiltonian perturbations, though the analysis becomes somewhat intricate. Guckenheimer and Holmes (1983) present this more general theory. It is clear that homoclinic crossings are structurally stable. A small perturbation of a Melnikov function with a nondegenerate zero produces another function with a zero.

3.2 The homoclinic tangle

Having established the fact that separatrices can intersect, let us deduce some of its consequences, and so build up the pattern of homoclinic motion. First we note that separatrices cannot cross themselves. This is because the image of an intersection is also an intersection. In the case of the stable manifold, there would be an infinity of self-intersections accumulating on the fixed point, in contradiction with the stable manifold theorem presented in section 2.2.

The images of the intersections of the stable manifold with the unstable manifold are again homoclinic points. Here there is no contradiction with the approximately linear structure near the fixed point (fig. 3.5). It should be noted that, in the absence of reflection about the fixed point (as in the weak perturbation of an equilibrium), the map orbits do not cross the separatrix. It follows that fig. 3.5 shows a pair of distinct homoclinic orbits arriving at the fixed point.

Another important property of a conservative map is that the successive loops formed between separatrix intersections must have the same area. But the distance between successive homoclinic points decreases exponentially as the fixed point is approached, so the 'length' of the loops must grow exponentially, while avoiding self-intersections. The only way out is

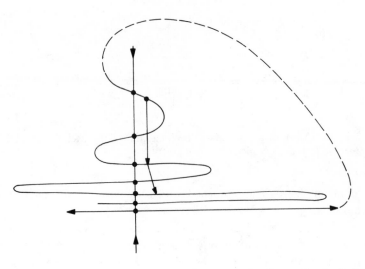

Fig. 3.5. Locally the orbits do not cross the separatrix. By continuity it follows that this figure shows a pair of alternate homoclinic orbits.

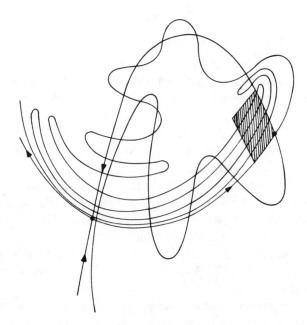

Fig. 3.6. The stable manifold has multiple intersections with the unstable manifold.

for the outgoing separatrix to accumulate back onto itself, as fig. 3.5 begins to show. This behaviour is in fact common to general (nonconservative) homoclinic motion, but the proof must then rely on the Palis lambda lemma (see Guckenheimer and Holmes, 1983).

Secondary intersections among the separatrices (fig. 3.6) are not avoided. There results an infinite number of distinct homoclinic orbits. In this context it is hard to resist the magic of Poincaré's own description (Poincaré, 1899), here roughly translated:

To even try to represent the figure formed by these two curves and their infinite intersections, each of which corresponds to a doubly asymptotic solution. The intersections form a kind of trellis, a tissue, an infinitely tight lattice; each of the curves must never self-intersect, but it must fold itself in a very complex way, so as to return and cut the lattice an infinite number of times.

The complexity of this figure is so astonishing that I cannot even attempt to draw it. Nothing is more appropriate to give us an idea of the complexity of the three-body problem and in general of all the problems in Dynamics where there is no uniform integral.

Let us consider the successive images of the small parallelogram in fig. 3.6. After five iterations of the map it will evolve into a long strip hugging the unstable manifold, so as to intersect the parallelogram (fig. 3.7). In general there will always be a kth iteration of the map, with the effect of stretching a neighbouring region of a homoclinic point and superposing it on the initial region. In the following section we will study the simplest example of such a process.

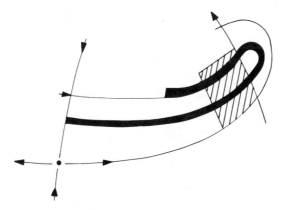

Fig. 3.7. The parallelogram shown in fig. 3.6 will become an elongated strip (with the same area) that eventually intersects its pre-image.

3.3 Symbolic orbits

Following the presentation of Lanford (1983), we seek a transformation **T** of the unit square R that stretches it and folds it over itself (fig. 3.8a). The mapping is not one-to-one: The set of points in R, which are brought back into R by **T**, consists of two vertical rectangles Δ_0 and Δ_1, and their images are two horizontal rectangles (fig. 3.8b). The simplest choice is to define the linear map

$$\mathbf{T}\begin{bmatrix} p \\ q \end{bmatrix} = \begin{bmatrix} \beta p + \gamma_{0,p} \\ \alpha q + \gamma_{0,q} \end{bmatrix} \text{ in } \Delta_0, \qquad \mathbf{T}\begin{bmatrix} p \\ q \end{bmatrix} = \begin{bmatrix} -\beta p + \gamma_{1,p} \\ -\alpha q + \gamma_{1,q} \end{bmatrix} \text{ in } \Delta_1 \quad (3.9)$$

with $\alpha > 1$ and $0 < \beta < 1$. Thus, **T** expands the square uniformly in the horizontal direction and contracts it along the vertical.

The problem is now to unravel the structure of the set of orbits that remain in R for all positive and negative iterations of **T**. Let $\mathbf{x} = (p, q)$ and $(\mathbf{x}_n, -\infty < n < \infty)$ be such an orbit. For $\mathbf{x}_{n+1} = \mathbf{T}\mathbf{x}_n$ to be in R, it is necessary that each $\mathbf{x}_n$ falls either in Δ_0 or in Δ_1. Since Δ_0 and Δ_1 have no points in common, $\mathbf{x}_n$ cannot be in both. Hence, there exists a unique index i_n (one or zero) such that $\mathbf{x}_n \in \Delta_{i_n}$. Proceeding in the same way for each n, we can associate a sequence $\{i_n\}$ of zeros and ones to each orbit, which remains in R for all n. The sequence $\{i_n\}$ is called the *symbolic description* of the orbit. We consider the sequence $\{i_n\}$ as a function of $\mathbf{x}_0$, $\{i_n(\mathbf{x}_0)\}$, also denoted by the (infinite) vector $\mathbf{i}(\mathbf{x}_0)$, the *symbolic orbit*. Evidently the orbit and its symbolic description are related by the *shift*

$$i_n(\mathbf{T}\mathbf{x}_0) = i_{n+1}(\mathbf{x}_0). \quad (3.10)$$

Can we invert the process; that is, is an orbit uniquely defined by its

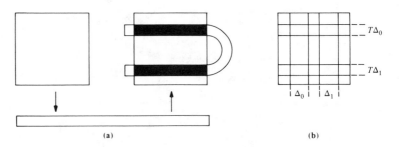

Fig. 3.8. (a) Smale's horseshoe map (3.9) linearly stretches the unit square and folds it back onto itself in imitation of homoclinic motion. (b) The images of the vertical strips Δ_0 and Δ_1 are horizontal strips.

symbol sequence? Can we associate an orbit to any arbitrary symbolic orbit? To settle these questions, we start by studying finite symbol sequences. Given the sequence $i_0, \ldots, i_{n-1}$ of zeros and ones, we define $\Delta_{i_0, \ldots, i_{n-1}}$ to be the set of all points that start out in Δ_{i_0}, then proceed to Δ_{i_1} and so on to $\Delta_{i_{n-1}}$. Using the fact that **T** expands horizontal separations by a factor α, we deduce that $\Delta_{i_0, \ldots, i_{n-1}}$ is a rectangle with unit height and width α^{-n}. Analogously, the set of points such that $T^j x_0 \in \Delta_{i_j}$ for $-n \le j \le -1$ is $T^n \Delta_{i_{-n}, \ldots, i_{-1}}$. This is another set of rectangles of unit length and height β^n. Each horizontal and vertical rectangle intersects in a little rectangle of height β^n and width α^{-n}, containing all the points for which $T^j x_0 \in \Delta_{i_j}$ for $-n \le j \le n-1$.

Figure 3.9 displays Δ_{i_0, i_1} and $\Delta_{i_{-2}, i_{-1}}$. In the limit $n \to \infty$ there results a sequence of inscribed rectangles, the sides of which tend to zero for any sequence **i**. Let the unique limiting point be x_0; it has the property that $T_j x_0 \in \Delta_{i_j}$ *for all j*, and no other point has this property. In short, we obtain a one-to-one correspondence between orbits and symbolic orbits.

The set of points contained in all the strips $\Delta_{i_0, \ldots, i_n}$ as $n \to \infty$ is an example of a *Cantor set*, also the case of the set $\Delta_{i_{-n}, \ldots, i_{-1}}$. Cantor sets are discussed within the general category of *fractals* in Mandelbrot (1982). Let

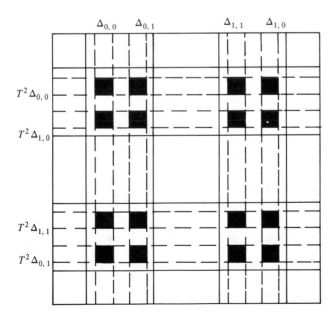

Fig. 3.9. The intersections of Δ_{i_0, i_1} and $\Delta_{i-2, i-1}$ are the sixteen black rectangles.

Λ be the set of all the points that never leave R; then the present and the future $(i_0, i_1, \dots)$ of the orbit of $x \in \Lambda$ determine its q coordinate, whereas its past is given by the coordinate p.

The unique correspondence between orbits and sequences of symbols permits us to infer the existence of orbits with definite types of recurrence, from the existence of sequences with the same property. For instance, a periodic orbit will be characterized by a periodic sequence of symbols. Therefore, Λ contains an infinite number of periodic orbits, corresponding to all the periodic symbolic orbits. In fact, it is easy to see that periodic orbits are dense in Λ: Take $x_0 \in \Lambda$ and let $\mathbf{i}$ be its symbolic orbit. We then define $\mathbf{i}'$ as the periodic sequence that coincides with $\mathbf{i}$ between the iterations $-n$ and $n + 1$. This sequence corresponds to the periodic point x_0', which lies in the same rectangle of sides α^{-n} and β^n as x_0, since they have the same symbolic sequence between $-n$ and $n + 1$. Choosing n arbitrarily large, we bring x_0' arbitrarily close to x_0.

There are 'well-behaved' orbits in Λ, but we also find chaotic motion. In particular, an orbit whose symbolic sequence contains all the finite sequences of zeros and ones comes arbitrarily close to all the points in Λ. In other words, there exist *dense orbits* in Λ.

The motivation for the study of Smale's horseshoe was its resemblance to high powers of the Poincaré map in the neighbourhood of homoclinic points. This analogy is made rigorous by the *Smale–Birkhoff homoclinic theorem* (proved in Guckenheimer and Holmes, 1983): Let T be a diffeomorphism, such that x_e is a hyperbolic fixed point and there exists a point $x \neq x_e$ of transverse intersection of the separatrices $W^+(x_e)$ with $W^-(x_e)$. Then T has a 'hyperbolic invariant set' Λ, on which T is topologically equivalent to a general shift on a symbolic orbit.

Two points should be explained about this theorem. The first is that it establishes the existence of a unique correspondence between the orbit in the invariant set Λ and a sequence of symbols., but this sequence can be more complex than the binary sequence in Smale's horseshoe; that is, there may be more 'rectangles' Δ_j than in this simple example. However, there must be points in each rectangle Δ_j that map into every one of the rectangles Δ_j'. The general shift equation is still given by (3.10).

It is also necessary to define what is meant by a 'hyperbolic' invariant set. We have already encountered hyperbolic motion in the case of the cat map (section 1.4). For a nonlinear map we define this property in terms of $D\mathbf{T}$, the derivative map introduced in section 2.1. Thus, a set Λ is called hyperbolic if we can decompose the tangent space of each point x in Λ into subspaces E_x such that

(a) if $v \in E_x^-$, then $|DT^n v| \leq C\gamma^n |v|$,

(b) if $v \in E_x^+$, then $|DT^{-n} v| \leq C\gamma^n |v|$,

where $C > 0$ and $0 < \gamma < 1$. In other words, for each point x in Λ there exists a plane E_x^- along which points infinitesimally close to x approach the orbit of x exponentially as $n \to \infty$, and another plane E_x^+ where the points approach the orbit of x as $n \to -\infty$.

It should be noted that, in the definition of hyperbolic set, the decomposition of $E^\pm$ is not presumed for the 'ambient space' surrounding Λ, though in the simple case of the cat map, Λ is the entire torus. The stable manifold theorem (section 2.2) can be generalized to hyperbolic sets; that is, there exist invariant manifolds $W_x^\pm$, tangent to $E_x^\pm$, whose points tend to the image $T^n x \in \Lambda$ as $n \to \mp \infty$. Of course, any periodic point in Λ is necessarily hyperbolic.

The immediate consequence of the Smale–Birkhoff theorem is that transverse homoclinic points lead to dense orbits and an infinite number of unstable periodic orbits, just as in the case of Smale's horseshoe. Horseshoe-type constructions have in fact been one of the main methods of establishing the existence of chaotic motion in dynamical systems. In particular, Devaney (1979) uses them to establish generic spiralling chaos in the neighbourhood of Hamiltonian equilibria with complex eigenvalues. The demonstration of Sitnikov's theorem for the three-body problem can also rely on the same method, as in Moser (1973). It must be pointed out, however, that horseshoes provide nonconstructive information. The calculation of the periodic orbits in Λ must rely on other methods, such as the one presented in chapter 4.

As we noted in section 3.1, homoclinic bifurcations also occur in dissipative systems. In their case the orbits converge on an attractor with smaller dimensionality than the phase space. This attractor is an invariant set; its coincidence with the hyperbolic set Λ is known as a *strange attractor*, discussed in Guckenheimer and Holmes (1983).

3.4 Measure and ergodicity

The goal of providing a complete qualitative description of generic dynamical systems is far from being achieved. It is therefore important to try to determine the main properties of 'almost all orbits' of a system. We adopt, then, a probabilistic point of view, which neglects 'exceptional trajectories' in favour of the 'typical' ones. To give a precise meaning to these vague concepts, we must fall back on the concept of *measure*. Abstract mathematical definitions of measure veil our familiarity with this term in the context of probability densities: If the (positive) probability density in a Euclidean space $\mathbb{R}^n$ is $f(\mathbf{x})$, then the *probability measure* in a region D is

$$\mu(D) = \int_D d\mu = \int_D d\mathbf{x}\, f(\mathbf{x}). \tag{3.11}$$

General measures are also defined by a positive density function, which does not have to be normalized to unity. Where a measure is defined on a set of discrete points, it is the sum of the measures for each point, in the same way as a probability.

An *invariant measure* for a map $\mathbf{T}$ is a measure with the property

$$\mu(\mathbf{T}(A)) = \mu(A) \tag{3.12}$$

for any set A. In sections 1.3 and 1.4 we used the same notation for the area of a region of the torus. This was justified, since the area itself was an invariant of the maps studied there. We also know that phase space volume is preserved by Hamiltonian flows; hence, the invariant measure for these systems will be the volume. The N-dimensional volume in $\mathbb{R}^N$ is known as the *Lebesgue measure*. If we focus on a single energy shell $H(\mathbf{x}) = E$, however, the preservation of volume on the whole phase space entails that the invariant measure corresponding to an element of area $d\sigma$ is the *Liouville measure*

$$d\mu = d\sigma/(\nabla H \cdot \nabla H)^{1/2} = \delta(E - H(\mathbf{x}))\, d\mathbf{x}. \tag{3.13}$$

Integrating over the shell of energy E and volume V we obtain the total measure $\mu(E) = dV/dE$. Other examples of invariant measure are (1) $\mu(\mathbf{x}_j) = 1/n$, in the case where the $\mathbf{x}_j$ form an n-fold periodic orbit; (2) $T(x) = 2x$ (mod 1) in the interval $(0, 1)$, which preserves the Lebesgue measure, though it is not a one-to-one mapping.

A fundamental result of the theory of invariant measures is the *Poincaré recurrence theorem*: Given a map $\mathbf{T}$ with an invariant probability measure μ and a measurable set A, if we define the set $A_0 = \{\mathbf{x} \in A : T^n\mathbf{x}$ eventually leave A permanently$\}$, then $\mu(A_0) = 0$.

Following Lanford (1983), we demonstrate this theorem by defining the set D_n of points whose image $T^n\mathbf{x} \in A$, while $T^j\mathbf{x} \notin A$ for $j > n$. Note that no requirement is made that $\mathbf{x}$ itself belong to A in this definition. Evidently the intersections $D_n \cap D_m = \varnothing$ (the empty set) if $n \neq m$. We also find that $\mathbf{x}$ is in D_n if and only if $T^n\mathbf{x}$ is in D_0, that is, $D_n = T^{-n}D_0$, which implies that

$$\mu(D_n) = \mu(D_0) \tag{3.14}$$

for all n. Therefore, the measure of the union of sets

$$\mu(D_0 \cup \cdots \cup D_{n-1}) = n\mu(D_0) \leq 1, \tag{3.15}$$

so that $\mu(D_0) = 0$, since n can be made arbitrarily large, and

$$\mu\left(\bigcup_{n=0}^{\infty} D_n\right) = 0. \tag{3.16}$$

The set A_0 in the theorem is a subset of this union set

$$A_0 = A \cap \left(\bigcup_{n=0}^{\infty} D_n\right). \tag{3.17}$$

Therefore, $\mu(A_0) = 0$. This theorem is also valid for a continuous flow.

A point $\mathbf{x}$ is a *wandering point* of the map $\mathbf{T}$ if there is a neighbourhood U of $\mathbf{x}$ and a number N for which $(\mathbf{T}^n U) \cap U = \varnothing$ for all $n > N$; otherwise, $\mathbf{x}$ is called a *nonwandering point*. If $\mathbf{T}$ is defined in a limited region of phase space or a compact manifold, the orbit of a wandering point must accumulate on the set of nonwandering points. Otherwise, the neighbourhoods of every point of the orbit, to which it cannot return, would eventually fill all the available phase space. It follows that the wandering points form an open set. Consider the union of these neighbourhoods U for all the wandering points. By the recurrence theorem, the measure of this set is zero, and hence the measure of the set of wandering points is also zero. In the case of a Hamiltonian system in which $H(\mathbf{x}) < E$ is a region with finite volume, the invariant measure is the volume itself. The set of wandering points is open and has zero volume; it is therefore empty. We conclude that in a Hamiltonian system with bounded motion all points are nonwandering.

We can consider now a function $f(\mathbf{x})$ defined on a region of phase space with an invariant measure μ. For *almost all points*, that is, with the exception of points forming a set of zero measure, the orbit will return again and again to the neighbourhoods already visited. If the intervals between these visits are not too erratic, we expect to be able to define the *time average*,

$$\bar{f}(\mathbf{x}) = N^{-1} \lim_{N \to \infty} \sum_{n=0}^{N-1} f(\mathbf{T}^n \mathbf{x}). \tag{3.18}$$

This is the main content of *Birkhoff's local ergodic theorem*, that is, the limit in (3.18) exists if $f(\mathbf{x})$ *is measurable*:

$$\int |f| d\mu < \infty. \tag{3.19}$$

This theorem also guarantees that $\bar{f}(\mathbf{T}\mathbf{x}) = \bar{f}(\mathbf{x})$ and that

$$\int \bar{f}(\mathbf{x}) \, d\mu = \int f(\mathbf{x}) \, d\mu \equiv \langle f(\mathbf{x}) \rangle. \tag{3.20}$$

Note that no guarantee is given that a given time average equals the *ensemble average* on the right side of (3.20). This will certainly not be the case for most periodic orbits, the time average of which is just the average over a single period. The fact that the set of points for which $\bar{f}(\mathbf{x})$ may not exist has zero measure is a strong restriction on bounded Hamiltonian systems. For dissipative systems, $\mu > 0$ only on attractors, outside of which no time average can be defined. Birkhoff's local ergodic theorem also holds for systems with continuous time. A proof is given in Billingsley (1965).

We have seen that in chaotic systems there are dense orbits. It is plausible that for one of these the *coincidence of averages*,

$$\bar{f}(\mathbf{x}) = \langle f(\mathbf{x}) \rangle, \tag{3.21}$$

will be verified. This is the motivation for the definition of *ergodic system* as one for which (3.21) holds for almost all orbits, for any measurable function f.

Exercise

Show that periodic orbits have zero measure in ergodic systems.

An important consequence of this definition is that it excludes any possibility of there existing two or more disjoint sets with positive measure in an ergodic system. For if two such sets, $\mathbf{T}M_1 = M_1$ and $\mathbf{T}M_2 = M_2$, existed, we could define the function

$$f(\mathbf{x}) = \begin{cases} 1 & (\mathbf{x} \in M_1) \\ 0 & (\mathbf{x} \in M_2) \end{cases} \tag{3.22}$$

with a time average that evidently depends on $\mathbf{x}$. Conversely, the system can always be decomposed if it is not ergodic. This follows from the fact that $\bar{f}(\mathbf{x})$ will not then be a constant for almost all $\mathbf{x}$. So we can find a suitable value a for which both the sets $M_1 = \{\mathbf{x} : \bar{f}(\mathbf{x}) < a\}$ and $M_2 = \{\mathbf{x} : \bar{f}(\mathbf{x}) > a\}$ have finite measure. According to Birkhoff's theorem the time average is invariant with respect to $\mathbf{T}$, so M_1 and M_2 are disjoint invariant sets with finite measure. Thus, a system is ergodic if and only if it cannot be decomposed; that is, any invariant set must have zero or unit (normalized) measure. Furthermore, an *invariant measurable function* $(f(\mathbf{T}\mathbf{x}) = f(\mathbf{x}))$ must be a constant for almost all $\mathbf{x}$.

Exercise

Show that a Hamiltonian system is never globally ergodic. What is referred to as an ergodic system in this case is the flow on the energy shell with the Liouville measure (3.13).

Another fundamental concept for the study of chaotic motion is that of 'mixing'. As motivation we take the classic example in Arnold and Avez (1968) – a glass containing 20 percent rum and 80 percent Coca Cola (fig. 3.10). If they are separated so that the region A is the region originally occupied by the rum and B is any region in the glass, after mixing n times the percentage of rum in B should be roughly

$$100\mu(\mathbf{T}^n A \cap B)/\mu(B) \simeq 100\mu(A) \tag{3.23}$$

if μ is normalized. In other words, we expect that each part of the glass will contain 20 percent of rum as $n \to \infty$. We therefore define a system to be *mixing* if

$$\lim_{n \to \infty} \mu(\mathbf{T}^n A \cap B) = \mu(A)\mu(B) \tag{3.24}$$

for any pair of measurable sets A and B.

The property of mixing implies ergodicity. To show this we choose an invariant measurable set A:

$$\mathbf{T}^n A \cap A = A \tag{3.25}$$

and the set $B = A$. Then (3.24) reduces to

$$\mu(A) = (\mu(A))^2, \tag{3.26}$$

wherefore $\mu(A) = 0$ or 1.

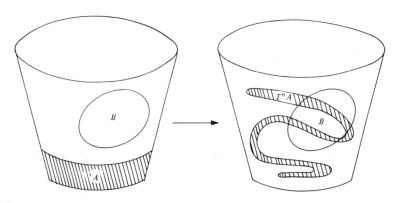

Fig. 3.10. Mixing 20% of rum into 80% of Coca Cola, there should be an equal proportion of rum in any region B of the glass.

Any region A will be distorted into a very long thin strip, winding helicoidally around the torus, by successive iterations of Arnold's cat map. This system is mixing and therefore it is also ergodic. As for the translations of the torus studied in section 1.3, they do not deform a region A. The intersections of $T^n A$ with a region B will alternate for all n between being empty, or having positive measure. This is an example of an ergodic system that is not mixing.

3.5 Entropy

A fundamental characteristic of chaotic motion is that any initial uncertainty about the system's position in phase space increases rapidly with time. Another way to put it is that successive approximate measurements of the system's position supply more information in a chaotic system than in a more 'predictable' system. The concept of metric entropy introduced by Kolmogorov turns this into a rigorous idea.

Let us make a partition α of phase space into N disjoint regions A_i, so that the probability of finding the system in any one of them is $\mu(A_i) = \mu_i$. We then define the *entropy of the partition* α to be

$$h(\alpha) = - \sum_{i=1}^{N} \mu_i \log \mu_i. \tag{3.27}$$

The partitions into N regions with the maximum entropy are the ones for which $\mu_i = 1/N$ for all i: $h = \log N$. This follows from the fact that the function $\phi(x) = -x \log x$ (fig. 3.11) is convex in the interval $[0, 1]$. Therefore,

$$\phi\left(N^{-1} \sum_{j=1}^{N} x_j \right) \geq N^{-1} \sum_{j=1}^{N} \phi(x_j) \tag{3.28}$$

and hence

$$h(\alpha) = \sum \phi(\mu_i) \leq N \phi(N^{-1} \sum \mu_i) = N \phi(N^{-1}) = \log N. \tag{3.29}$$

The trivial partition, where one of the regions has unit measure, is the only one with zero entropy.

The probability of finding the system in a region A_i of the partition α may or may not be independent of the probability of finding it in the region B_j of the partition β. For example, a mixing system that was in B_j to start with will certainly be in $T^n B_j$ – scattered throughout phase space after a large number n of iterations. In this case the probabilities of the partitions β and $\alpha \equiv \lim T^n \beta$, for $n \to \infty$, are independent. The combined partition into

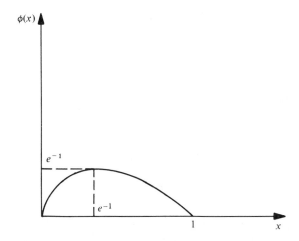

Fig. 3.11. The convexity of $\phi(x) = -x \log x$ is fundamental for the proof of important properties of the metric entropy.

the regions $A_i \cap B_j$ with probabilities $\mu_{ij} = \mu(A_i \cap B_j)$ is denoted $\alpha \vee \beta$. If α and β are independent, $\mu_{ij} = \mu_i \mu_j$.

Exercise

Show that, if α and β are independent, the entropy of the combined partition is

$$h(\alpha \vee \beta) = h(\alpha) + h(\beta). \tag{3.30}$$

When α and β are not independent, we define the conditional probability of finding the system in A_i, given that it is in B_j, as

$$\mu_{i/j} = \mu_{ij}/\mu_j. \tag{3.31}$$

The combined entropy then becomes

$$h(\alpha \vee \beta) = -\sum_{ij} \mu_j \mu_{i/j} (\log \mu_j + \log \mu_{i/j})$$

$$= -\sum_j \mu_j \log \mu_j \sum_i \mu_{i/j} - \sum_j \mu_j \sum_i \mu_{i/j} \log \mu_{i/j}. \tag{3.32}$$

Defining the *conditional entropy of α relative to B_j* as

$$h(\alpha/B_j) = -\sum_i \mu_{i/j} \log \mu_{i/j} \tag{3.33}$$

and the *conditional entropy of α relative to the partition β* as the mean value for $h(\alpha/B_j)$,

$$h(\alpha/\beta) = \sum_j \mu_j h(\alpha/B_j), \qquad (3.34)$$

we obtain

$$h(\alpha \vee \beta) = h(\beta) + h(\alpha/\beta). \qquad (3.35)$$

Exercise

Use the convexity of the function in fig. 3.11 to prove that

$$h(\alpha/\beta) \le h(\alpha). \qquad (3.36)$$

We can now return to the question of incomplete information about the system and its evolution. If a measurement localizes the system in one of the regions A_j, we reduce the uncertainty concerning its localization. The information produced by the measurement consists of the removal of uncertainty that preceded this event. The larger the uncertainty, the greater will be the amount of information obtained by its removal. The entropy of a partition α has been shown to have properties that permit us to consider it as a measure of the uncertainty of α. We can therefore define the *quantity of information* obtained by localizing the system in one of the regions A_j as $h(\alpha)$.

If we make a measurement on α, having already ascertained the B_j into which the system belongs in the partition β, the average increase of information will be $h(\alpha/\beta)$. This additional information is a maximum if α and β are independent – in this case a measurement of β in no way reduces the uncertainty of α. Though we have found that the present definition of entropy has desirable properties, it may still appear somewhat arbitrary. However, it is shown in Khinchin (1957) that this is the only definition that satisfies conditions (3.29) and (3.35).

We focus on the sequence of partitions $T^n \alpha$. It is demonstrated in Arnold and Avez (1968) that the limit

$$h(\alpha, \mathbf{T}) \equiv \lim_{n \to \infty} \frac{h(\alpha \vee \mathbf{T}\alpha \vee \cdots \vee \mathbf{T}^{n-1}\alpha)}{n} \qquad (3.37)$$

exists. We call $h(\alpha, \mathbf{T})$ the entropy of α relative to $\mathbf{T}$. Finally we define the *metric entropy* of the map $\mathbf{T}$ as

$$h(\mathbf{T}) \equiv \sup h(\alpha, \mathbf{T}), \qquad (3.38)$$

where the supremum extends over all measurable partions. Evidently we always have $h(\mathbf{T}) \ge 0$.

Exercise

Prove that the metric entropy is invariant with respect to a coordinate transformation $x' = F(x)$, that is, $h(T) = h(F \cdot T \cdot F^{-1})$.

It is not usually feasible to compute the metric entropy of a dynamical system directly from its definition. Yet it is worthwhile to try to gain a feeling for the process involved. Each iteration leads to a new combined partition, which results from the addition of the boundaries of the regions $T^n A_j$ to the boundaries of all the previous partitions. For example, if the original partition divided phase space into $2N$ regions of equal measure and each iteration of the map led to a new partition that cut each region in half, then $h(\alpha \vee \cdots \vee T^{n-1}\alpha) = n \log 2 + \log N$, so $h(\alpha, T) = \log 2$. This example shows that positive entropy requires exponential growth, with n, of the number of regions in $\alpha \vee \cdots \vee T^{n-1}\alpha$. But it must be remembered that the number of regions in $T^n\alpha$ is the same as that in α, so an exponential growth of the combined partition demands an exponential growth of the boundary of the regions $T^n A_j$. Only thus can the boundaries of $T^n\alpha$ cut a constant proportion of the ever diminishing regions $\alpha \vee \cdots \vee T^{n-1}\alpha$. This does not occur for the translations of the torus studied in section 1.3. The length of the boundary of any region $T^n A_j$ is constant; hence, the entropy $h(\alpha, T)$ is zero for any partition α, even though this system is ergodic.

For the cat map we do have exponential stretching of the boundaries of each region along the direction of the eigenvector ξ_1 corresponding to the eigenvalue λ_1. If each region were collapsed back on itself as was the case with the horseshoe (but fitting exactly, as the full cat map itself), each region would be divided into λ_1 equal regions. The result would be that

$$h(T) = \log \lambda_1. \tag{3.39}$$

This is the correct result, proved by Sinai (1959), which implies that the existence of many extra regions (due to the boundaries not fitting properly) is counterbalanced by the fact that the entropy is actually smaller than the maximum $\log N$. This result can be generalized to r-dimensional tori: if a linear map T of the torus has r real and distinct eigenvalues, then

$$h(T) = \sum_{|\lambda_i| > 1} \log|\lambda_i|. \tag{3.40}$$

It is usually very difficult to verify whether the properties of ergodicity, mixing and positive entropy hold for a given Hamiltonian system. The first systems for which this has been achieved are *billiards*, that is, the free motion of a particle in the plane within a given boundary. Usually one

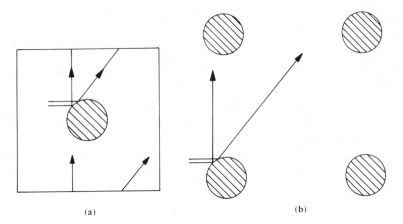

Fig. 3.12. In Sinai's billiard the particle undergoes specular reflection at the circular barrier, while the boundary conditions on the square are periodic. Thus, the square in (a) is in fact a torus with the periodic representation depicted in (b).

considers that the particle undergoes specular reflection at the boundary, but in *Sinai's billiard* shown in fig. 3.12a this is true only for encounters with the circular barrier. The boundary conditions along the square are taken to be periodic, so that the billiard in fact has the topology of a torus. The periodic representation of the torus is shown in fig. 3.12b. This system was proved by Sinai (1970) to have all the chaotic properties, including positive entropy. Another system proved to be chaotic by Buminovitch (1974) is the *stadium*: the interior of a pair of semicircles joined by two straight segments.

The concept of entropy permits us to quantify the decay of information brought about by the entire flow of the system. Yet the loss of memory about approximate initial conditions is a local property of chaotic motion. Let us consider a small ball of radius $\varepsilon(0)$ centred on the initial position of some orbit in phase space. The successive images of the ball will be distorted under the map **T**. In the limit $\varepsilon(0) \to 0$, we can use the linearized map $D\mathbf{T}$, so that the ball develops into a small ellipsoid. Naming the principal axes of the ellipsoid as $\varepsilon_i(n)$, we define the *Lyapunov exponents* as

$$\gamma_i = \lim_{\varepsilon(0) \to 0} \lim_{n \to \infty} \frac{1}{n} \log \frac{\varepsilon_i(n)}{\varepsilon(0)}. \tag{3.41}$$

Exercise

Define the Lyapunov exponents in terms of the tangent space vectors introduced in chapter 2.

For a periodic orbit of period r, the above limit evidently exists and the γ_i are just the eigenvalues of $\mathcal{M}^r = D\mathbf{T}^r$. Where periodic orbits are dense, we therefore expect that most orbits will have Lyapunov exponents. This is indeed the content of a theorem of Oseledec (1968), which holds under very general conditions. Furthermore, for a hyperbolic set with dense orbits, almost all orbits will share the same Lyapunov exponents. The largest positive exponent measures the rate of divergence between neighbouring orbits, whereas negative exponents measure convergence rates along stable manifolds. For a Hamiltonian system $\sum \gamma_i = 1$, whereas $\sum \gamma_i < 1$ for dissipative systems.

It was shown in the preceding section that the metric entropy will be positive only if the frontiers of a typical region grow exponentially with time. The possibility of choosing such a region as the small ball used to define the Lyapunov exponent suggests that positive entropy is connected to positive Lyapunov exponents. In the case of linear maps of the torus we simply have $\gamma_i = \ln|\lambda_i|$, so equation (3.40) becomes

$$h(\mathbf{T}) = \sum_{\gamma_i > 1} \gamma_i. \tag{3.42}$$

Pesin (1977) has indeed shown that (3.42) can be generalized for a single region of connected chaotic motion. If there is more than one region (separated by invariant curves or manifolds), the generalization is

$$h(\mathbf{T}) = \int d\mu \sum_{\gamma_i > 1} \gamma_i(\mathbf{x}). \tag{3.43}$$

This connection between the Lyapunov exponents and the metric entropy is of great practical importance, since the latter are much easier to compute. Lichtenberg and Lieberman (1983, chap. 5) present a detailed discussion of this subject and provide many references.

3.6 The principle of uniformity

Hyperbolic sets are characterized by dense orbits and by a dense population of periodic orbits. The ergodic theory, briefly outlined in the two preceding sections, relies on the former and neglects the latter. Yet a dense orbit mimics every periodic orbit of the system arbitrarily closely for many periods, so it should be possible to replace the time average over a 'typical orbit' by an average over all the periodic orbits. Clearly the chaotic orbit will remain close to very unstable periodic orbits for a shorter time than the relatively stable ones. It follows that the mean over periodic orbits must be a weighted average.

The principle of uniform density of periodic orbits is most easily presented in the case of an area-preserving map in a region of area A, as in the appendix of Hannay and Ozorio de Almeida (1984). Let us consider a peaked function $\delta_\varepsilon(\mathbf{x})$, centred on the origin but normalized to unity for all ε, so that for $\varepsilon \to 0$ it becomes a two-dimensional Dirac δ function. Then $\delta_\varepsilon(\mathbf{x}_n - \mathbf{x}_0)$ is a function of $\mathbf{x}_0$ defined throughout the area A, which has peaks in all the regions where orbits return very close to their starting positions after precisely n iterations. Another function of $\mathbf{x}_0$ that we can construct is

$$\Delta_{\varepsilon,N}(\mathbf{x}_0) = \frac{1}{N} \sum_{n=1}^{N} \delta_\varepsilon(\mathbf{x}_n - \mathbf{x}_0), \tag{3.44}$$

peaked in all the regions with orbits that return to a small neighbourhood of $\mathbf{x}_0$ for some iteration $n \geq N$.

One of the consequences of the Poincaré recurrence theorem discussed in section 3.4 is that all points of an area-preserving map are nonwandering; that is, the image of any neighbourhood U of x_0 will eventually intersect U. If we take this neighbourhood to be the area ε around $\mathbf{x}_0$, we see that there will be a peak of $\Delta_{\varepsilon,N}$ in the neighbourhood of any point $\mathbf{x}_0$ in the limit $N \to \infty$. This uniform distribution of peaks will make $\Delta_{\varepsilon,N}(\mathbf{x}_0)$ a smooth function over A. In fact, we can estimate when this will happen: Defining $v(N)$ to be the number of periodic points with period equal to or smaller than N, the number of peaks in $\Delta_{\varepsilon,N}$ will be equal to or greater than $v(N)$ (equality being obtained in the limit $\varepsilon \to 0$). The smoothing of $\Delta_{\varepsilon,N}$ will then occur when

$$\varepsilon v(N) \gg A. \tag{3.45}$$

For an ergodic system we can readily calculate this smooth limit, since

$$\lim_{N \to \infty} \Delta_{\varepsilon,N}(\mathbf{x}_0) = \lim_{N \to \infty} \frac{1}{N} \sum_{n=1}^{N} \delta_\varepsilon(\mathbf{x}_n - \mathbf{x}_0) = \bar{\delta}_\varepsilon(\mathbf{x}_n - \mathbf{x})_{\mathbf{x}_0 = \mathbf{x}}. \tag{3.46}$$

It is important to note that the 'time' average on the right side is now taken over a function of $\mathbf{x}_0$, centred on the immobile point $\mathbf{x}$. The limit exists for almost all $\mathbf{x}_0$ and in particular for $\mathbf{x}_0 = \mathbf{x}$. The ergodic property then becomes

$$\lim_{N \to \infty} \Delta_{\varepsilon,N}(\mathbf{x}_0) = \langle \delta_\varepsilon(\mathbf{x}_0 - \mathbf{x}) \rangle = A^{-1}. \tag{3.47}$$

The *principle of uniformity* can now be stated as the conjecture that we can take the limit $\varepsilon \to 0$ on the left side of (3.47) before the $N \to \infty$ limit. In practice this means that we can make the ε neighbourhoods sufficiently small so as to be able to neglect the contribution of nonperiodic orbits to $\Delta_{\varepsilon,N}$, while making $v(N)$ large enough for (3.45) to be valid. In conclusion

$$\lim_{N\to\infty} \frac{1}{N} \sum_{n=1}^{N} \delta(\mathbf{x}_n - \mathbf{x}_0) = \frac{1}{A} \qquad (3.48)$$

is a sum over periodic orbits, though the equation makes sense only for arbitrarily small smoothing of the δ functions. Integrating both sides of (3.48) over $\mathbf{x}_0$, we obtain the *sum rule over periodic points*,

$$\lim_{N\to\infty} N^{-1} \sum_{n=1}^{N} \sum_{j=1}^{v(n)-v(n-1)} |\det(\mathcal{M}_j^n - 1)|^{-1} \to 1. \qquad (3.49)$$

The weight of a periodic point is given by a factor that appeared in (2.23). It depends on the stability matrix $\mathcal{M}$ introduced in section 2.3 and is unbounded as the linearized map $D\mathbf{T}^n$ approaches the unit map.

For a fully chaotic system with positive entropy $h(\mathbf{T})$ the weight of an orbit of very high period will have the limit

$$|\det(\mathcal{M}_j^n - 1)|^{-1} \xrightarrow[n\to\infty]{} |(e^{n\gamma} - 1)(e^{-n\gamma} - 1)|^{-1} \xrightarrow[n\to\infty]{} e^{-n\gamma} = \exp[-nh(\mathbf{T})],$$

$$(3.50)$$

where γ is the positive Lyapunov exponent. The contribution of any finite number of orbits to the sum in (3.49) can be neglected, so we can use (3.50) for all the orbits. Since the contribution of each orbit is $n \exp(-nh)$, we obtain the average number of periodic orbits with period n as

$$v(n) - v(n-1) \xrightarrow[n\to\infty]{} \exp(nh)/n. \qquad (3.51)$$

The above argument is strictly valid only if n is a prime number. Otherwise, we have neglected the repeated contribution of orbits for which n/m is an integer for some integer m. However, taking (3.51) as a first approximation we verify that the number of orbits with period n/m is indeed exponentially smaller than the number with period n. Rigourous results concerning uniformity and exponential proliferation of periodic orbits are presented by Parry and Pollicot (1983) and Parry (1984).

Exercise

Though the uniformity principle is restricted to ergodic systems, show that Birkhoff's local ergodic theorem can be used to extend the validity of the sum rule (3.49) to systems that are merely recurrent.

The uniformity principle and the sum rule may be applied to the Poincaré map of a Hamiltonian system. It is worthwhile, however, to extend them to the continuous periodic orbits in the flow itself. The first difficulty is that ergodic flow on the energy shell does not preserve area, though it does conserve the Liouville measure (3.13). For this reason we define the *shell δ function* $\delta_\mu(\mathbf{x})$ by its properties

$$\delta_\mu(\mathbf{x}) = 0, \qquad \mathbf{x} \neq 0 \tag{3.52}$$

(choosing the origin to lie on the shell) and

$$\int d\mu\, \delta_\mu(\mathbf{x}) = 1. \tag{3.53}$$

Just as with the usual Dirac δ functions, we must understand this to be the limit of peaked functions, but I will leave out explicit reference to the smearing parameter so as not to encumber the notation.

The time average can be written in the form

$$\bar{\delta}_\mu(\mathbf{x}(t) - \mathbf{x}) = \lim_{T \to \infty} \frac{1}{2T} \int_{-T}^{T} dt\, \bar{\delta}_\mu(\mathbf{x}(t) - \mathbf{x}), \tag{3.54}$$

where the slash in the integral sign excludes an arbitrarily small neighbourhood of the origin. Its equivalence to the configurational average

$$\langle \delta_\mu(\mathbf{x}(0) - \mathbf{x}) \rangle = \mu^{-1} = (dV/dE)^{-1}, \tag{3.55}$$

for the choice of $\mathbf{x}(0) = \mathbf{x}$, leads to the uniformity in $\mathbf{x}(0)$ of

$$\lim_{T \to \infty} \frac{1}{2T} \int_{-T}^{T} dt\, \bar{\delta}_\mu(\mathbf{x}(t) - \mathbf{x}(0)) = \mu^{-1}. \tag{3.56}$$

This formula for the uniform distribution of periodic orbits is exactly analogous to the one for maps. The need to exclude a neighbourhood of the origin from the average over time here becomes evident.

The sum rule over periodic orbits is obtained by integrating over the measure μ of the whole shell:

$$\lim_{T \to \infty} \frac{1}{2T} \int_{-T}^{T} dt \int d\mu\, \delta_\mu(\mathbf{x}(t) - \mathbf{x}(0)) = 1. \tag{3.57}$$

The configurational integral can be subdivided into integrals in the neighbourhood of each periodic orbit. For each of these we can introduce the local coordinate system presented in section 2.5. The shell δ function δ_μ then takes on the simple form

$$\delta_\mu(\theta, P, Q) = (dH/dJ)\, \delta(\theta, P, Q). \tag{3.58}$$

Consequently,

$$
\int_{j\text{th orbit}} dt\, d\mu\, \delta_\mu(\mathbf{x}(t) - \mathbf{x}(0)) = \int dt\, d\theta_0\, dP_0\, dQ_0\, \delta(\theta_0 + \omega_j t - (\theta_0 + 2\pi m))
$$
$$
\times \delta((P,Q)_t - (P,Q)_0)
$$
$$
= |\tau_j|\, |\det\{\mathscr{M}_j^m - \mathbf{1}\}|^{-1}, \tag{3.59}
$$

so that the sum rule becomes

$$
\lim_{T\to\infty} \frac{1}{2T} \sum_{|m\tau_j|<T} |\tau_j|\, |\det\{\mathscr{M}_j^m - \mathbf{1}\}|^{-1} = 1. \tag{3.60}
$$

As in the discrete case, it is not necessary to presuppose ergodicity to arrive at this sum rule, though it is essential to the uniformity principle (3.56). For a system with positive entropy we can neglect the contribution of multiple repetitions of periodic orbits, just as in the discrete case. For large periods the number of periodic orbits increases exponentially, permitting us to define the *weighted density of periodic orbits*,

$$
f(\tau) = \sum_j |\tau_j|\, |\det\{\mathscr{M}_j - \mathbf{1}\}|^{-1}\delta(\tau - \tau_j) \xrightarrow[\tau\to\infty]{} 1, \tag{3.61}
$$

as can be verified by integrating over τ.

In chapter 9, which presents the theory of the quantum energy spectrum in terms of the classical periodic orbits, the factor $|\det(\mathscr{M}^m - \mathbf{1})|^{-1}$ will reappear and (3.61) becomes the basis for an important result concerning correlations of the spectrum.

4

Normal forms

Having established the ubiquity of periodic orbits in dynamical systems, we now return to the study of the motion near a given periodic orbit, fixed point or point of equilibrium. The Hartman–Grobman theorem of section 2.2 guarantees the existence of a continuous coordinate transformation that linearizes the vector field near a hyperbolic fixed point, but no indication is given as to how to construct this transformation. The method of normal forms, invented by Poincaré, consists of eliminating nonlinear terms of the vector field by successive polynomial transformations. If this process can be carried out to all orders, the resulting compound transformation can be shown to be convergent in some cases, and an *analytic* reduction of the nonlinear vector field to a linear one is thus achieved. This transformation can be approximated to arbitrary accuracy.

One of the cases in which this process can never be carried out is that of Hamiltonian systems. The Hamiltonian cannot generally be made quadratic by a canonical transformation, though Birkhoff showed that it can be simplified into a form that shares some of the important features of quadratic Hamiltonians. For hyperbolic points this transformation is analytic in a narrow neighbourhood of the separatrices, allowing us to calculate precisely some homoclinic orbits and the periodic orbit families that accumulate on them.

The elimination of nonlinear terms of low order near stable points in Hamiltonian systems allows a qualitative understanding of the motion in their neighbourhood, even though the transformation cannot be extended to infinite orders. This is important in the case of the bifurcation of stable orbits discussed at the end of this chapter. The nonconvergence of the Birkhoff transformation is due to the intricate nature of the motion in these regions, which is studied in chapters 5 and 6.

4.1 General systems

A linear coordinate transformation can take a dynamical system into the form

$$\dot{\mathbf{x}} = \mathscr{L}\mathbf{x} + \mathbf{g}_k(\mathbf{x}) + \mathbf{g}_{k+1}(\mathbf{x}) + \cdots \tag{4.1}$$

in the neighbourhood of the origin, an equilibrium with N distinct eigenvalues. Here $\mathscr{L}$ is a diagonal matrix with elements λ_j, and the components of the vector fields $\mathbf{g}_j$ are homogeneous polynomials of order j. We seek a nonlinear coordinate transformation

$$\mathbf{x} = \mathbf{y} + \mathbf{h}(\mathbf{y}), \tag{4.2}$$

the components of $\mathbf{h}$ being homogeneous polynomials of order $\mathbf{k}$, that eliminates $\mathbf{g}_k$ from the dynamical system. Introducing (4.2) into (4.1), we obtain

$$\dot{\mathbf{y}} = (1 + \partial\mathbf{h}/\partial\mathbf{y})^{-1}\{\mathscr{L}(\mathbf{y} + \mathbf{h}(\mathbf{y})) + \mathbf{g}_k(\mathbf{y} + \mathbf{h}(\mathbf{y})) + \cdots\}$$
$$= \mathscr{L}\mathbf{y} + \{\mathscr{L}\mathbf{h}(\mathbf{y}) - (\partial\mathbf{h}/\partial\mathbf{y})\mathscr{L}\mathbf{y} + \mathbf{g}_k(\mathbf{y})\} + \cdots, \tag{4.3}$$

neglecting all terms of order $\geq k + 1$. The required polynomials $h_j(\mathbf{y})$ thus satisfy the equations

$$\lambda_j h_j - \sum_i \frac{\partial h_j}{\partial y_i} \lambda_i y_i = g_{kj}. \tag{4.4}$$

The substitution of h_j by any of its monomials $(y_1^{m_1} \cdots y_N^{m_N})$ on the left side of (4.4) gives

$$\lambda_j(y_1^{m_1} \cdots y_N^{m_N}) - \sum_i \lambda_i y_i \frac{\partial}{\partial y_i}(y_1^{m_1} \cdots y_N^{m_N}) = \left(\lambda_j - \sum_i \lambda_i m_i\right)(y_1^{m_1} \cdots y_N^{m_N}); \tag{4.5}$$

so we can solve (4.4) by choosing the coefficients of the monomials in h_j to be equal to the coefficients in g_{kj} divided by $(\sum_i \lambda_i m_i - \lambda_j)$.

The transformation eliminating the terms $\mathbf{g}_k(\mathbf{x})$ from the vector field modifies the higher-order terms, but the form of the dynamical system will be the same as (4.1), with k replaced by $k + 1$. The process of elimination can be carried through indefinitely unless there exists a vector $\mathbf{m}$ with positive integer components, for which

$$\lambda_j = \sum_{i=1}^N m_i \lambda_i = \mathbf{m} \cdot \mathscr{L} \tag{4.6}$$

for one of the eigenvalues. The equality (4.6) is known as the *resonance condition*, and the smallest $k = \sum_i m_i$ for which it holds is the *order of the resonance*. We can obviously eliminate all the terms $\mathbf{g}_2 \cdots \mathbf{g}_{k-1}$ of a resonant system of order k, and if the system is nonresonant, we can formally eliminate all the nonlinear terms. Examples of resonant systems are those with $\lambda_1 = 2\lambda_2$ (second order) and $\lambda_1 + \lambda_2 = 0$ (third order, since $\lambda_1 = 2\lambda_1 + \lambda_2$). The relation $2\lambda_1 = 3\lambda_2$ does not imply a resonance.

It is easy to correlate some resonances to the configuration of the motion. This is the case of a conjugate pair of imaginary eigenvalues. As we have seen in section 1.2, the linear motion then has a centre, so that the nonlinear terms may push the orbits either towards the origin or away from it. The case of two degenerate real (negative) eigenvalues can likewise be understood. The linear motion is then radial towards the origin. This is the borderline between two nodes, having either eigenvector as the 'fast' or the 'slow' axis. A perturbation may also make the dynamical matrix undiagonalizable.

Example. Consider the system $\dot{\mathbf{x}} = \mathbf{f}(x)$:

$$\dot{x}_1 = \lambda_1 x_1, \qquad \dot{x}_2 = \lambda_2 x_2 + a x_1^3. \tag{4.7}$$

The transformation $\mathbf{x} \rightarrow \mathbf{y}$,

$$y_1 = x_1^3, \qquad y_2 = x_2, \tag{4.8}$$

linearizes this system in the form

$$\dot{y}_1 = 3\lambda_1 y_1, \qquad \dot{y}_2 = \lambda_2 y_2 + a y_1. \tag{4.9}$$

This system can be diagonalized by the transformation $\mathbf{y} \rightarrow \mathbf{Y}$:

$$Y_1 = y_1, \qquad Y_2 = y_2 - \frac{a}{3\lambda_1 - \lambda_2} y_2. \tag{4.10}$$

Finally, the transformation $\mathbf{Y} \rightarrow \mathbf{\symbf{x}}$,

$$Y_1 = \symbf{x}_1^3, \qquad Y_2 = \symbf{x}_2, \tag{4.11}$$

brings the dynamical system to the form $\dot{\symbf{x}} = D\mathbf{f}(\symbf{x})$. The full transformation $\symbf{x} \rightarrow \mathbf{x}$ is simply

$$x_1 = \symbf{x}_1, \qquad x_2 = \symbf{x}_2 + \frac{a}{3\lambda_1 - \lambda_2} \symbf{x}_1^3. \tag{4.12}$$

It is analytic everywhere unless there is a resonance: $\lambda_2 = 3\lambda_1$. In this case there is nothing special about the resonant system itself. However, it becomes impossible to diagonalize the degenerate linear system (4.10).

Exercise

It is possible to reduce systems with degenerate eigenvalues to a normal form if the eigenvalues are degenerate, though the solution no longer decouples in terms of monomials (Arnold, 1982). Find the linear equations for the coefficients in **h** for $k = 2$ if the diagonal matrix can be reduced only to a two-dimensional Jordan block.

The absence of an exact resonance does not guarantee the convergence of the *normal-form transformation* to the resulting linear system. The problem is that the resonance condition may be approached closely many times with arbitrary precision, as $|\mathbf{m}| \to \infty$. This is known as the *small-denominator problem*: Some of the coefficients of the successive normal-form transformations become arbitrarily large. The necessary and sufficient condition for this not to occur is that $\mathscr{L}$ lie in the *Poincaré domain*: All the eigenvalues lie on one side of a straight line on the complex λ plane, whereas the origin lies on the other side. To see that this is sufficient, we project all the eigenvalues onto the normal to the straight line, as in fig. 4.1. All of these will be greater than or equal to the distance between the straight line and the origin. Hence, the linear combination of eigenvalues $\mathbf{m} \cdot \mathscr{L}$ has an arbitrarily large normal projection as $|\mathbf{m}| \to \infty$. It must therefore become greater than the projection of any individual eigenvalue.

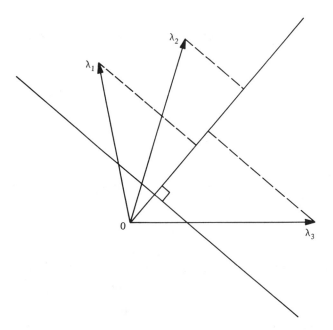

Fig. 4.1. The eigenvalues lie in the Poincaré domain if they are all on the opposite side of a straight line with respect to the origin. Convergence of the normal form in some neighbourhood of the origin is guaranteed in this case.

Poincaré demonstrated the convergence of the normal-form transformation in a neighbourhood of the equilibrium for eigenvalues in the Poincaré domain. In the *Siegel domain*, complementary to the Poincaré domain, convergence is still possible according to *Siegel's theorem* (demonstrated in Arnold, 1982, sec. 28). The sufficient condition is that there exist positive numbers C and v such that

$$|\lambda_j - \mathbf{m} \cdot \mathscr{L}| \geq C/|\mathbf{m}|^v \tag{4.13}$$

for each eigenvalue λ_j and for all positive-integer vectors $\mathbf{m}$. This strange condition resembles the one for the convergence of the circle map, to be discussed in detail in the next chapter.

There are two alternatives by which we can extend the normal-form analysis to periodic orbits. One is to consider the Poincaré section near a periodic orbit and to derive the normal form in the neighbourhood of the fixed point. This procedure, presented in Arnold (1982), is simple for general systems, but its adaptation to Hamiltonian systems is not easy. Otherwise, we can use periodic coordinates based on the periodic orbit, as in section 2.5, and work with the reduced time-dependent system.

Consider, then, the linear part of this time periodic system:

$$\dot{\mathbf{x}} = \mathscr{A}(t)\mathbf{x}. \tag{4.14}$$

We now define the matrix $\mathscr{L}$ so that the map after the period 2π takes the form

$$\mathbf{x} = \exp(2\pi\mathscr{L})\mathbf{x}_0 = \left(\exp \int_0^{2\pi} \mathscr{A}(t)\,dt \right)\mathbf{x}_0. \tag{4.15}$$

The coordinate transformation

$$\mathbf{x} = \left(\exp \int_0^t \mathscr{A}(t)\,dt \right) \exp(-t\mathscr{L})\mathbf{y} \tag{4.16}$$

can then be verified to reduce (4.14) to the time-dependent linear system

$$\dot{\mathbf{y}} = \mathscr{L}\mathbf{y}. \tag{4.17}$$

If $\mathscr{L}$ has distinct eigenvalues, it can be diagonalized, so we will suppose $\mathscr{L}$ to be diagonal as in the time-independent theory.

We seek again a nonlinear coordinate transformation that eliminates the term $\mathbf{g}_k$ (of order ≥ 2) from the expansion

$$\dot{\mathbf{x}} = \mathscr{L}\mathbf{x} + \mathbf{g}_k(\mathbf{x}, t) + \mathbf{g}_{k+1}(\mathbf{x}, t) + \cdots. \tag{4.18}$$

Thus substitution

$$\mathbf{x} = \mathbf{y} + \mathbf{h}(\mathbf{y}, t), \tag{4.19}$$

where **h** is a vector with polynomial components h_j in **y** that has periodic coefficients in t, leads to the set of equations

$$\lambda_j h_j - \sum_i \frac{\partial h_j}{\partial y_i} \lambda_i y_i - \frac{\partial h_j}{\partial t} = -g_{kj}(\mathbf{y}, t). \tag{4.20}$$

Expanding the components in Fourier series,

$$h_j(\mathbf{y}, t) = \sum_{l=-\infty}^{\infty} \sum_{m_1 + \cdots + m_N = k} h_{j,m,l} y_1^{m_1} \cdots y_N^{m_N} e^{ilt}, \tag{4.21}$$

and proceeding likewise with the field components g_{kj}, we obtain the formal solution

$$h_{j,m,l} = \frac{g_{kj,m,l}}{il + \mathbf{m} \cdot \mathcal{L} - \lambda_j}, \tag{4.22}$$

unless the resonance condition

$$\lambda_j = \mathbf{m} \cdot \mathcal{L} + il \tag{4.23}$$

is satisfied.

The question of convergence of the normal-form series is dealt with in much the same way as in the time-dependent problem. The difference lies in the imaginary number il, which limits the Poincaré domain to be the set of matrices $\mathcal{L}$, the eigenvalues of which have all positive real parts or all negative real parts. Convergence is guaranteed in the Poincaré domain, whereas small denominators again appear in the complementary Siegel domain.

4.2 The Birkhoff normal form

It was shown in section 1.2 that the eigenvalues of Hamiltonian systems always arise in pairs: $\lambda, -\lambda$. In the context of normal forms they therefore have a third-order resonance. Hence, it is impossible to linearize Hamiltonian systems by an analytic canonical transformation. The best that one can aim for is to simplify their Taylor expansion at an equilibrium or periodic orbit, as with non-Hamiltonian resonances treated by Arnold (1982) and Guckenheimer and Holmes (1983).

The special feature of the Birkhoff normal form is that it is a reduction of the Hamiltonian rather than the full dynamical system. The *generating function* $S(\mathbf{p}, \mathbf{q}, t)$ defines implicitly a canonical transformation $(\mathbf{p}, \mathbf{q}) \to (\mathbf{p}, \mathbf{q})$ by the equations

$$\mathbf{p} = \partial S / \partial \mathbf{q}, \qquad \mathbf{q} = \partial S / \partial \mathbf{p}. \tag{4.24}$$

The Hamiltonian is concomitantly transformed by

$$H(\mathbf{p}, \mathbf{q}) = H(\mathbf{p}, \mathbf{q}) + \partial S/\partial t. \tag{4.25}$$

The idea is thus to study the effect of polynomial generating functions on the Taylor expansion of the Hamiltonian. The treatment here will be limited to a single degree of freedom, to avoid encumbering the notation, for the generalization is obvious. Since the system is time periodic, the foregoing also applies to the neighbourhood of a periodic orbit of an autonomous system with two freedoms, as discussed in section 2.5.

The simplest case is the unstable equilibrium. The Taylor–Fourier expansion of the Hamiltonian then has the form

$$H(p, q, t) = \lambda pq + \sum_{\alpha+\beta=k}^{\infty} \sum_{l=-\infty}^{\infty} H_{\alpha\beta l} p^\alpha q^\beta \exp(ilt), \tag{4.26}$$

where $k > 2$. This is transformed by the generating function

$$S(\mathbf{p}, q, t) = \mathbf{p}q + \sum_{\alpha+\beta=k} \sum_{l=-\infty}^{\infty} S_{\alpha\beta l} \mathbf{p}^\alpha q^\beta \exp(ilt). \tag{4.27}$$

According to (4.24) the first term of S generates the identity transformation. The second term will be small, of order k, near the origin. The series for the first two partial derivatives,

$$p = \partial S/\partial q = \mathbf{p} \left[1 + \sum_{l,\alpha+\beta=k} \beta S_{\alpha\beta l} \mathbf{p}^{\alpha-1} q^{\beta-1} \exp(ilt) \right],$$

$$\mathbf{q} = \partial S/\partial \mathbf{p} = q \left[1 + \sum_{l,\alpha+\beta=k} \alpha S_{\alpha\beta l} \mathbf{p}^{\alpha-1} q^{\beta-1} \exp(ilt) \right],$$

$$\partial S/\partial t = \sum_{l,\alpha+\beta=k} il S_{\alpha\beta l} \mathbf{p}^\alpha q^\beta \exp(ilt), \tag{4.28}$$

can be inverted up to order k to yield the explicit transformation,

$$p = \mathbf{p} \left[1 + \sum_{l,\alpha+\beta=k} \beta S_{\alpha\beta l} \mathbf{p}^{\alpha-1} q^{\beta-1} \exp(ilt) + \cdots \right],$$

$$q = \mathbf{q} \left[1 - \sum_{l,\alpha+\beta=k} \alpha S_{\alpha\beta l} \mathbf{p}^{\alpha-1} q^{\beta-1} \exp(ilt) + \cdots \right],$$

$$\frac{\partial S}{\partial t} = \sum_{l,\alpha+\beta=k} il S_{\alpha\beta l} \mathbf{p}^\alpha q^\beta \exp(ilt) + \cdots. \tag{4.29}$$

The quadratic part of the Hamiltonian becomes

$$\lambda pq = \lambda \mathbf{p}q \left[1 + \sum_{l,\alpha+\beta=k} (\beta-\alpha) \mathbf{p}^{\alpha-1} q^{\beta-1} \exp(ilt) + \cdots \right], \tag{4.30}$$

so the full Hamiltonian takes the form

$$H(\mathfrak{p}, \mathfrak{q}, t) = \lambda \mathfrak{p} \mathfrak{q} + \sum_{l, \alpha + \beta = k} \{H_{\alpha\beta l} + [\lambda(\beta - \alpha) - il] S_{\alpha\beta l} \mathfrak{p}^\alpha \mathfrak{q}^\beta \exp(ilt) + \cdots\}.$$

(4.31)

We can therefore eliminate all the terms of order k with the choice

$$S_{\alpha\beta l} = \frac{-H_{\alpha\beta l}}{\lambda(\beta - \alpha) - il},$$

(4.32)

except for the term $\alpha = \beta$, $l = 0$, that is, the term $H_{\alpha\alpha 0}(\mathfrak{p}\mathfrak{q})^\alpha$. Note that the inclusion of such a term in (4.26), with $2\alpha < k$, would not alter in any way the form of the k-order terms in (4.31). Hence, we can iterate the process of eliminating higher and higher order terms of the Hamiltonian, so as to obtain the kth-order *Birkhoff normal form*

$$H(\mathfrak{p}, \mathfrak{q}, t) = \lambda \mathfrak{p} \mathfrak{q} + H_2(\mathfrak{p}\mathfrak{q})^2 + \cdots + H_k(\mathfrak{p}\mathfrak{q})^{k/2} + \sum_{l, \alpha + \beta = k+1} H_{\alpha\beta l} \mathfrak{p}^\alpha \mathfrak{q}^\beta \exp(ilt)$$

(4.33)

for any even k.

Differentiating the Hamiltonian in the normal form results in a polynomial dynamical system with odd-order terms, which correspond to the resonant terms of the Poincaré normal form discussed in the previous section. Within that context the Hamiltonian system is in the Siegel domain; that is, in addition to the exact resonances, it is also subject to small denominators. The nontrivial content of Birkhoff's construction is seen from the fact that there are no small denominators in the successive transformations whose generating functions have coefficients given by (4.32). The expectation that the normal-form transformation is convergent for $k \to \infty$ is indeed borne out by Moser (1956) for a disc around the origin. We shall discuss the motion near an unstable orbit in the next section.

The Hamiltonian near a stable equilibrium has the Taylor-Fourier expansion

$$H(\mathfrak{p}, \mathfrak{q}, t) = \frac{\omega}{2}(\mathfrak{p}^2 + \mathfrak{q}^2) + \sum_{\alpha + \beta = 3}^{\alpha + \beta = \infty} \sum_l H'_{\alpha\beta l} \mathfrak{p}^\alpha \mathfrak{q}^\beta \exp(ilt),$$

(4.34)

where $k > 2$. This form is not convenient for the deduction of the normal-form transformation, so we complexify the Hamiltonian; that is, considering p and q to be complex variables, we make the linear transformation

$$z = p + iq, \qquad z^* = p - iq.$$

(4.35)

The variables z and z^* are considered to be independent. However, if the initial conditions for p and q are real, $z(t)$ and $z^*(t)$ will remain complex conjugate solutions for all t, as discussed in section 1.1.

Exercise

Verify that the trajectories $z(t)$ and $z^*(t)$ satisfy Hamilton's equations with the Hamiltonian

$$-2iH(p(z,z^*),q(z,z^*),t) = -i\omega zz^* + \sum_{\alpha+\beta=3}^{\alpha+\beta=\infty} \sum_l H_{\alpha\beta l} z^\alpha z^{*\beta} \exp(ilt).$$

(4.36)

Following exactly the same method as for the unstable case, we can now reduce the Hamiltonian to the kth-order normal form:

$$-2iH = -i\omega\mathfrak{z}\mathfrak{z}^* + H_2(\mathfrak{z}\mathfrak{z}^*)^2 + \cdots + H_k(\mathfrak{z}\mathfrak{z}^*)^{k/2}$$
$$+ \sum_{l,\alpha+\beta=k+1} H_{\alpha\beta l}\mathfrak{z}^\alpha\mathfrak{z}^{*\beta} \exp(ilt).$$

(4.37)

That is, we eliminate the 'nondiagonal' coefficients of the Hamiltonian through the choice

$$S_{\alpha\beta l} = \frac{iH_{\alpha\beta l}}{\omega(\alpha-\beta)-l}$$

(4.38)

for the coefficients of the generating function

$$S(\mathfrak{z},z^*,t) = \mathfrak{z}z^* + \sum_{l,\alpha+\beta=k} S_{\alpha\beta l}\mathfrak{z}^\alpha z^{*\beta} \exp(ilt)$$

(4.39)

of the transformation $(z,z^*) \to (\mathfrak{z},\mathfrak{z}^*)$.

This process can be carried through indefinitely only if the frequency ω is not a rational number. Otherwise, the maximum order for the normal form will be the highest $\alpha+\beta=k$ for which the *resonance condition*

$$\omega(\alpha-\beta)-l=0$$

(4.40)

is not met. Even for irrational frequencies we are beset by small denominators, arising for all α, β and l, such that $l(\alpha-\beta)$ is a good rational approximation to ω. These near resonances lead usually to the divergence of the normal form for stable periodic orbits, as will be shown in section 4.4. Only the time-independent Hamiltonian with a single freedom is exempt from resonances or small denominators.

The procedure for deducing the normal form for autonomous systems with two freedoms (or more) is essentially the same as above. If the linearized Hamiltonian is $\lambda_1 p_1 q_1 + \lambda_2 p_2 q_2$, we can use the real coordinates, obtaining a normal form with small denominators (and resonances if λ_1 and λ_2 are rationally related). For $\lambda_1 p_1 q_1 + (\omega_2/2)(p_2^2 + q_2^2)$, we proceed by complexifying the (p_2, q_2) coordinates. The resultant normal form has no small denominators. Finally, we complexify both pairs of coordinates to obtain the normal form for a stable equilibrium. This is the case that is studied most extensively in spite of the presence of small denominators. Most presentations follow a more abstract algebraic approach, which leads to useful computational methods. These are reviewed by Lichtenberg and Lieberman (1983).

Exercise

Show that the complexification

$$p_1' = p_1 - ip_2, \qquad q_1' = q_1 + iq_2$$
$$p_2' = p_1 + ip_2, \qquad q_2' = q_1 - iq_2 \tag{4.41}$$

takes the linear Hamiltonian system $H = \lambda(p_1 q_1 + p_2 q_2) + \omega(p_1 q_2 - q_1 p_2)$, with a complex quartet of eigenvalues, into the form $H' = (\lambda - i\omega)p_1' q_1' + (\lambda + i\omega)p_2' q_2'$. Show that the normal form for H' has no small denominators.

4.3 Homoclinic motion revisited

The *Birkhoff–Moser theorem*, which establishes the existence of an analytic transformation reducing an unstable time-periodic system to its Birkhoff normal form, also applies to maps. This is obvious for the Poincaré map of the system, and Moser (1956) proves that one can always interpolate a time-periodic Hamiltonian system, given an area-preserving map. The normal form of the map will simply be the Poincaré map of the normal form for the interpolated time-periodic system. Since the normalized Hamiltonian is independent of time and depends only on the product pq: $H(p, q, t) = f(\xi)$, where $\xi = pq$, Hamilton's equations have the form

$$\dot{p} = -(df/d\xi)p, \qquad \dot{q} = (df/d\xi)q. \tag{4.42}$$

Verifying that ξ and hence $df/d\xi$ are constants of the motion, we can immediately integrate (4.42) and so obtain the Poincaré map for a period of 2π, $\mathbf{x}' = \mathbf{T}_N(\mathbf{x})$, as

$$p' = U(pq)p, \qquad q' = U^{-1}(pq)q, \tag{4.43}$$

where

$$U(\xi) = \exp(-i2\pi df/d\xi). \tag{4.44}$$

For a time-periodic system we get a different normal-form transformation for each Poincaré map by fixing t in (4.26) and (4.27). Of course, it is not necessary to go through the business of interpolating systems once the existence of the normal form for maps has been proved. All that must be done is to evaluate the coefficients of

$$U(\mathfrak{p}\mathfrak{q}) = \lambda \left[1 + \sum_{\alpha=1}^{\infty} U_{2\alpha}(\mathfrak{p}\mathfrak{q})^{\alpha} \right] \tag{4.45}$$

and the normal-form transformation $\mathbf{x} = \mathbf{N}(\mathbf{\mathfrak{x}})$

$$p = \mathfrak{p} + \sum_{\alpha+\beta=2}^{\infty} p_{\alpha\beta}\mathfrak{p}^{\alpha}\mathfrak{q}^{\beta},$$

$$q = \mathfrak{q} + \sum_{\alpha+\beta=2}^{\infty} q_{\alpha\beta}\mathfrak{p}^{\alpha}\mathfrak{q}^{\beta}, \tag{4.46}$$

by direct substitution into the map $\mathbf{x}' = \mathbf{T}(\mathbf{x})$:

$$p' = \lambda p + \sum_{\alpha+\beta=2}^{\infty} p'_{\alpha\beta}p^{\alpha}q^{\beta},$$

$$q' = \lambda^{-1}q + \sum_{\alpha+\beta=2}^{\infty} q'_{\alpha\beta}p^{\alpha}q^{\beta}. \tag{4.47}$$

Moser (1956) proves the convergence of the normal-form transformation in a small disc surrounding the origin. This is not of immediate use, because the disc is an unstable region, only briefly visited by most points. It is therefore important to show that the region of convergence can be extended, as in da Silva Ritter, Ozorio de Almeida and Douady (1987), for an analytic map whose inverse $\mathbf{T}^{-1}$ is also analytic. This is possible because the normal map (4.43) is defined in terms of the function $U(\mathfrak{p}\mathfrak{q})$, which is itself invariant under the mapping. If the series for the normal map converges for $(\mathfrak{p}_0, \mathfrak{q}_0)$ in D_0, it converges for all $(\mathfrak{p}, \mathfrak{q})$ such that $\mathfrak{p}\mathfrak{q} \leq \mathfrak{p}_0\mathfrak{q}_0$. The relation between the invariant curves and the region of convergence D_0 and $\mathbf{N}(D_0)$ for the original and the normalized system is shown in fig. 4.2.

Let us now define the transformations

$$\mathbf{N}_m = \mathbf{T}^{-m} \cdot \mathbf{N} \cdot \mathbf{T}_N^m, \qquad \mathbf{N}_0 = \mathbf{N} \tag{4.48}$$

for all positive or negative m. Since $\mathbf{N}$ is analytic in D_0, $\mathbf{N}_{-m}$ is analytic in the region

$$D_m = \mathbf{T}_N^m(D_0). \tag{4.49}$$

All the regions D_n have points in common, namely, some neighbourhood

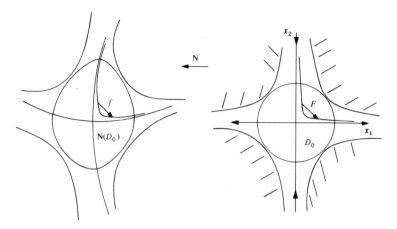

Fig. 4.2. The circle of convergence D_0 and the invariant hyperbolas that touch it are distorted by the normal-form transformation **N**.

of the origin (fig. 4.3). Moreover, $\mathbf{N}_0$ and $\mathbf{N}_m$ coincide in the region where they both converge:

$$\mathbf{N}_m = \mathbf{T}^{-m} \cdot \mathbf{N} \cdot \mathbf{T}_N^m = \mathbf{N} \cdot \mathbf{T}_N^m \cdot \mathbf{N}^{-1} \cdot \mathbf{N} \cdot \mathbf{T}^m = \mathbf{N}. \tag{4.50}$$

The same is true for any pair of regions D_m and $D_{m'}$, so we can define the extended normal-form transformation to be

$$\mathbb{N} = \lim_{K \to \infty} \bigcup_{m=-K}^{K} \mathbf{N}_m, \tag{4.51}$$

which will be analytic in the region

$$\mathbb{D} = \lim_{K \to \infty} \bigcup_{m=-K}^{K} D_m. \tag{4.52}$$

Because of (4.50), the Taylor series for each $\mathbf{N}_m$ will be identical with the series for $\mathbf{N}_0$, but we must show that the region of convergence for the Taylor series of $\mathbf{N}_m$ indeed extends far out of D_0. This is done in da Silva Ritter, Ozorio de Almeida and Douady (1987) by complexifying, that is, by using the fact that the convergence of the series for $\mathbf{N}_0(\mathbf{x}_0)$ implies the convergence of the Taylor series for $\mathbf{N}_0(\mathbf{\jmath})$, for $|\mathbf{\jmath}_1| \leq |\mathbf{x}_{10}|$ and $|\mathbf{\jmath}_2| \leq |\mathbf{x}_{20}|$. Therefore, inside this region there are no poles, nor are there any in the complex regions D_{nc} obtained from it by the transformations (4.48). This fact in turn ensures the convergence of the Taylor series in D_{nc}.

The separatrices of the map **T** are the images of the $\mathbf{x}_1 = 0$ and $\mathbf{x}_2 = 0$

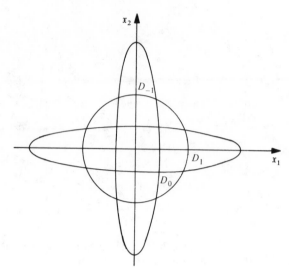

Fig. 4.3. All the images D_n of D_0 have a common intersection, which includes the origin.

axes under the transformation $\mathbb{N}(\mathbf{x})$. The extension of the Birkhoff–Moser theorem guarantees that the normal form converges, indefinitely far out along the separatrices. As a consequence we can calculate homoclinic (and heteroclinic) points from the images of both axes under $\mathbb{N}$. (Heteroclinic points are calculated from the normal forms for two unstable points.) Though it is not difficult to compute some homoclinic points directly by following orbits along the stable and unstable manifolds, it is only the above method that provides the entire homoclinic orbit with uniform precision. All that has to be done is to follow the linear orbit of the normal coordinates for the homoclinic point, before transforming back to $\mathbf{x}$. This last step is not even necessary close to the origin, since there $\mathbb{N}$ is asymptotic to the identity.

The normal-form transformation also converges along invariant square hyperbolas very close to the $\mathbf{x}_1 = 0$ and $\mathbf{x}_2 = 0$ axes. The images of these curves self-intersect near the homoclinic points. These self-intersections are not generally periodic points, but a simple argument due to Birkhoff (1927) shows that they give rise to an infinite sequence of periodic points. Consider the orbit of a point of self-intersection; generally there will be $n - 1$ images along the invariant curve, say the inner one in fig. 4.4a, while the nth image lies beyond the self-intersection. However, the average spacing between the points on the orbit diminishes continuously as the

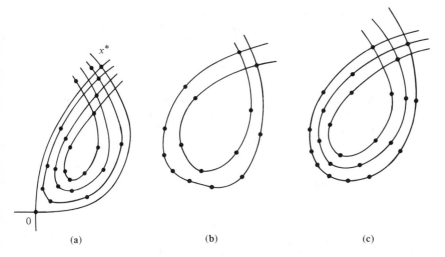

Fig. 4.4. (a) The three self-intersecting curves within the stable and the unstable manifolds are the images of invariant hyperbolas. There are five images of the point of intersection on the inner curve and six on the outer curve. The boundary curve supports a periodic orbit of period 6. (b) Here the pair of curves supports a single periodic orbit of period 15. (c) A periodic orbit of period 24 is supported on three open invariant curves.

invariant curves come closer to the separatrices, that is, for self-intersections that approach the homoclinic point. There will thus be an invariant curve where the nth image of the self-intersection comes back to itself exactly – the nth-order periodic orbit. By the same token we find an $(n + 1)$-order periodic orbit and so on.

There are many other families of periodic points that accumulate on the homoclinic point in the limit of infinite periods. For instance, there are the families based on the mutual intersections of a pair of invariant curves shown in fig. 4.4b. These are obtained by requiring that two families of curves satisfy two conditions: The orbit of each one of the intersections must land exactly on the other intersection. This is also the situation for the simplest heteroclinic points. Figure 4.4c sketches a periodic orbit based on three invariant curves. In general we can find periodic orbits bound to any number k of invariant curves. In the limit where $k \to \infty$, we obtain recurrent chaotic orbits.

The larger the period of the orbit, the closer will be the hyperbolas on which they travel to the x axes. In this limit the map $\mathbf{T}_N$ becomes linear. Since the periodic points accumulate on the homoclinic point, their position is accurately given by a linearization of the normal-form transformation around the homoclinic point. Combining both linearizations, we arrive at explicit linear equations for the periodic points, which are asymptotically exact in the limit of high periods. To achieve this, let us consider both images $(x_1^*, 0)$ and $(0, x_2^*)$ of the homoclinic point x^* under the transformation N^{-1} (fig. 4.5). We define N_1 and N_2 to be the restrictions of N to neighbourhoods of these two points. A periodic point of order m, with normal coordinates x_m, propagates from near $(0, x_2^*)$ by T_N^m into the neighbourhood of $(x_1^*, 0)$. It satisfies the equation

$$x_m = N_2^{-1} \cdot N_1 \cdot T_N^m(x_m). \tag{4.53}$$

Linearizing N_j around the images of the homoclinic point and the normal map T_N^m:

$$\begin{bmatrix} x_1' \\ x_2' \end{bmatrix} = \mathscr{L}^m x = \begin{bmatrix} \lambda & 0 \\ 0 & \lambda^{-1} \end{bmatrix}^m x, \tag{4.54}$$

we can approximate (4.53) by the linear equation

$$\mathscr{R}^m x \equiv D[N_2^{-1} \cdot N_1 \cdot T_N^m] x = (0, x_2^*) + (DN_2)^{-1}(DN_1)[\mathscr{L}^m x - (x_1^*, 0)]. \tag{4.55}$$

A periodic orbit $x_{m_J \cdots m_1}$ with m_1 points on one hyperbola, m_2 on another and so on up to m_J on the Jth hyperbola is obtained from the solution of the equation

$$\left(\prod_{j=1}^J \mathscr{R}_{m_J} - 1 \right) x_{m_J \cdots m_1} = 0. \tag{4.56}$$

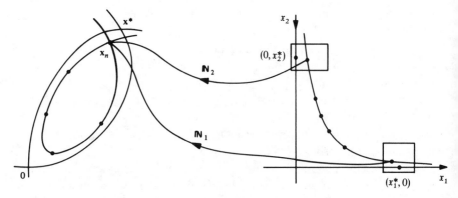

Fig. 4.5. $N_1(x)$ and $N_2(x)$ are, respectively, the restrictions of $N(x)$ to the neighbourhoods of the pre-images of the homoclinic point $(x_1^*, 0)$ and $(0, x_2^*)$.

Contrary to appearances, (4.56) is not a homogeneous equation, the inhomogeneous terms arising from the normal coordinates for the homoclinic point in (4.55). Thus, we obtain a unique solution for any set of indices $m_1 \cdots m_J$. The resulting approximation for $x_{m_J \cdots m_1}$ becomes asymptotically exact in the limit where all the $m_J \to \infty$.

It is interesting that the presence of denumerable periodic orbits is a consequence of the Smale–Birkhoff theorem and ultimately derives from the nonlinearity of the map $\mathbf{T}$. It may seen strange to calculate these orbits from a linear map, but it must be borne in mind that the linearization is performed around the homoclinic point and not the origin. We are familiar with chaotic maps where the orbits are 'folded back in'. This is the case of Arnold's cat map, which also has denumerable periodic orbits, and even for Smale's horseshoe, developed as a simplified model of homoclinic motion.

The computational feasibility of calculating homoclinic orbits from the Birkhoff normal form was verified by Ozorio de Almeida, Coutinho and da Silva Ritter (1985) for the family of quadratic maps

$$x_1' = x_1 \cosh \alpha + (x_2 - x_1^2) \sinh \alpha,$$
$$x_2' = x_2 \sinh \alpha + (x_2 - x_1^2) \cosh \alpha. \tag{4.57}$$

This can be proved to be equivalent to all quadratic maps with a hyperbolic fixed point at the origin, just as the similar map used by Henon (1969) is equivalent to all quadratic maps containing an elliptic fixed point. Figures 4.6a and b show, respectively, the separatrices calculated by iterating points on the separatrices close to the origin and by the normal form. Da Silva Ritter et al. (1987) achieved the calculation of unstable orbits that returned to themselves after ten iterations within an accuracy of twenty-one significant figures, by calculating the self-intersection of invariant curves! Good agreement was found with the periodic points calculated from the explicit linear equation (4.56) (with $J = 1$), and the accuracy of the points calculated in this way was seen to improve with n.

This technique can evidently be extended to the calculation of hetero-clinic points and their satellite periodic orbits and to the calculation of homoclinic (heteroclinic) points of time-periodic Hamiltonian systems with one freedom. Though the numerical convergence of the normal form is very fast, the very large x coordinates needed to calculate 'higher homoclinic points' arising from the mutual intersections of long loops of the separatrices may not be accessible. It seems likely that the present formalism can be used to compute homoclinic and periodic orbits near unstable equilibria for which the normal form has no small denominators.

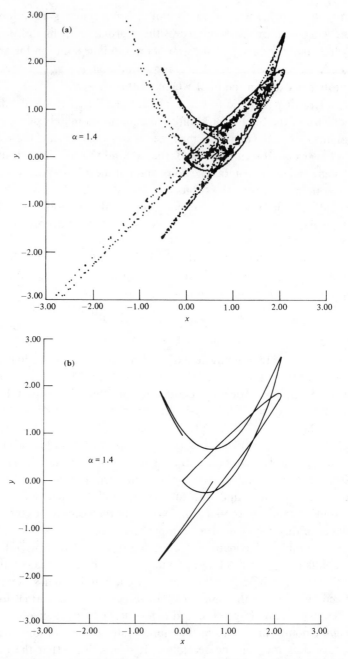

Fig. 4.6. (a) Points on the stable and unstable manifolds obtained by positive and negative iterations on the linear approximation of the manifolds very near the origin, for $\alpha = 1.4$. (b) The two separatrices calculated from the Birkhoff normal form up to twentieth order.

4.4 Bifurcations of stable periodic orbits

In the absence of an exact resonance, the cancelling of all the 'off-diagonal' terms in the normal form for a stable periodic orbit leads to the formal series

$$-2iH = -i\omega zz^* + H_2(zz^*)^2 + \cdots. \tag{4.58}$$

We can cast this into real variables by merely reversing the transformation (4.35), so that

$$H = \omega\left(\frac{p^2 + q^2}{2}\right) + H_2\left(\frac{p^2 + q^2}{2}\right)^2 + \cdots, \tag{4.59}$$

or we can use the canonical polar coordinates (1.48), that is, action-angle variables,

$$p = (2I)^{1/2}\sin\phi, \qquad q = (2I)^{1/2}\cos\phi, \tag{4.60}$$

to obtain

$$H = \omega I + H_2 I^2 + \cdots. \tag{4.61}$$

The Hamiltonian not only is integrable, because of the loss of its time dependence, but also is independent of θ. The circles centred on the origin are invariant circles, just as with the quadratic Hamiltonian $\omega I = \omega(p^2 + q^2)/2$. The difference is that the motion along each circle no longer has the universal angular velocity ω. Indeed, from Hamilton's equations

$$\dot{\phi}(I) = \omega + 2H_2 I + \cdots. \tag{4.62}$$

Even though the normal form does not usually converge, we can bring any nonresonant Hamiltonian to coincide with it in its lowest terms. The normal form therefore provides an accurate approximation of the motion close to the origin, that is, the periodic orbit of an autonomous two-freedom system. In this expanded picture (fig. 4.7), the invariant circle of the normal form corresponds to a thin torus surrounding the periodic orbit. Neighbouring orbits wind around such tori. The frequency of rotation $\dot{\phi}(I)$ varies continuously, so that arbitrarily close to the origin there are Birkhoff tori for which $\dot{\phi} = l/k$ is a rational number. The orbits on these tori are periodic, for the orbits of the Poincaré map $t \rightarrow t + 2\pi$ close after k iterations.

Let us focus on one of these *periodic tori*, while altering the energy. It was shown in section 2.5 that typical periodic orbits are members of one-parameter families, where the energy can be chosen as the parameter. Generally the linear frequency will be a monotonic function $\omega(E)$ over a small energy range. So if the periodic orbit torus lies close to the periodic orbit at the orign, we may choose instead

$$\varepsilon = \omega(E) - l/k \tag{4.63}$$

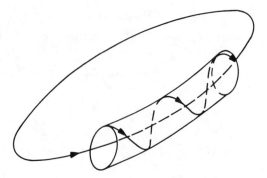

Fig. 4.7. The invariant circle of the normal form corresponds to a thin torus surrounding the periodic orbit.

as the parameter for the family. Substituting l/k on the left side of (4.62) and taking ω from (4.63), the action variable for the periodic torus is seen to be

$$I_{l/k} = -\varepsilon/(2H_2) + O(\varepsilon^2). \qquad (4.64)$$

Thus, the periodic torus collapses onto the origin as $\varepsilon \to 0$, which is exactly the resonance condition. This event resembles the generic Hopf bifurcations of dissipative systems described by Arnold (1982) and Guckenheimer and Holmes (1983). The orbits on the periodic torus are fixed points of the kth iteration of the Poincaré map. According to the 'Birkhoff picture', this entire family of fixed points collapses onto the central fixed point when a resonance occurs. This is a complete alteration of the 'linear picture', where at resonance all points become fixed points. The resonance condition for the Poincaré map is the same as (2.33).

As the resonance condition is approached, one of the denominators in the normal-form transformation becomes arbitrarily large. The Birkhoff normal form does not hold through the resonance, so we must work instead with a *resonant normal form,* including all the terms that blow up the transformation exactly at the resonance. This has the form

$$-2iH = -i\omega zz^* + H_2(zz^*)^2 + \cdots + H_{k0l}z^k e^{ilt} + H_{0k,-l}z^{*k}e^{-ilt} + \cdots, \qquad (4.65)$$

showing explicitly the resonant term of lowest order, since this will have the strongest effect near the origin, where the bifurcation takes place. Though the resonant terms cannot be cancelled when $\varepsilon = 0$, we can still eliminate the time dependence by means of the canonical time-dependent transformation

$$\zeta = z \exp(ilt/k), \qquad \zeta^* = z^* \exp(-ilt/k), \tag{4.66}$$

generated implicitly by

$$\sigma(z, \zeta^*, t) = z\zeta^* \exp(ilt/k). \tag{4.67}$$

Applying (4.25), we obtain

$$-2iH = -i\varepsilon\zeta\zeta^* + H_2(\zeta\zeta^*)^2 + \cdots + 2i\,\mathrm{Im}(H_{k0l}\zeta^k) + \cdots, \tag{4.68}$$

using the fact that $H_{0k,-m} = -H_{k0m}^*$ for H to be real. In the real polar coordinates

$$\zeta = (2I)^{1/2} \exp(i\phi), \tag{4.69}$$

the Hamiltonian takes the form

$$H(I, \phi) = \varepsilon I + H_2 I^2 + \cdots + aI^{k/2} \sin(k\phi). \tag{4.70}$$

So the coordinate transformation (4.66) takes the system to rotating coordinates in which the periodic torus (4.64) becomes a circle of fixed points. For $k > 4$ the last term, as well as the higher ϕ independent terms, are smaller than the first two as $I \to 0$. We can then consider the effect of the ϕ dependent term as a perturbation. Because of it, in general $(dH/d\phi)_{I_{l/k}} \neq 0$, so that the only fixed points on the periodic torus have $\phi = (\pi/2 + n\pi)/k$. As a first approximation around the unperturbed torus $I_{l/k}$, we have

$$H(I, \phi) = H_2(I - I_{l/k})^2 + aI_{l/k}^{k/2} \sin k\phi, \tag{4.71}$$

which can be recognized as the Hamiltonian for the pendulum; its level curves are shown in fig. 4.8. So one-half of the periodic points are stable and one-half unstable. The stable points are surrounded by closed invariant curves known as *islands* or *island tori* in the extended two-freedom system. These correspond to librations of the pendulum. The pendulum rotations are distortions of the unperturbed Birkhoff invariant curves [horizontal lines, in the (I, ϕ) coordinates]. Between rotations and librations lie the separatrices. Since the system is time independent, we obtain their equation from the level curve with the energy of the unstable fixed point:

$$I - I_{l/k} = [aI_{l/k}^{k/2}(1 - \sin k\phi)/H_2]^{1/2}. \tag{4.72}$$

Therefore, the maximum area of the islands is of order $\varepsilon^{k/4}$. For $k > 4$ the islands shrink faster than the periodic tori as $\varepsilon \to 0$. Thus, the 'Birkhoff picture', though qualitatively incorrect right down to $\varepsilon \to 0$, becomes

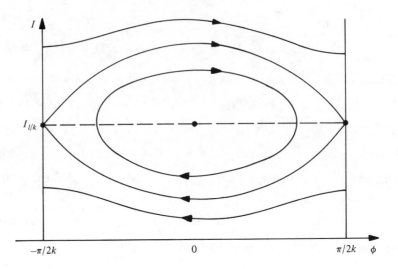

Fig. 4.8. The Hamiltonian for the resonant normal form is approximately the same as for a pendulum with angular momentum $I - I_{l/k}$.

quantitatively accurate: It is impossible to distinguish a very thin resonance from a torus in a computation.

The foregoing analysis focused on a single periodic Birkhoff torus followed through the exact resonance by means of a survey of the family of periodic orbits placed at the origin. For a single energy E there is an infinite number of periodic Birkhoff tori, the frequency of which approximates the irrational frequency $\omega(E)$ arbitrarily well. The above theory can be applied to every one of them. In each case the periodic Birkhoff torus breaks up into $2k$ periodic points, corresponding to a single pair of stable and unstable periodic orbits in the extended system. These resonances are dense near the origin, so the Birkhoff normal form cannot converge onto the wrong picture of the motion.

We note that even the resonant normal form does not converge. To start with, it displays only a single resonance, so it is wrong in the region of rotations of the pendulum Hamiltonian (4.62). To the truncated resonant normal form, we must add the time-dependent remainder. We therefore expect that the separatrices will split into the transverse homoclinic intersections studied in chapter 3. Yet the closer the periodic Birkhoff torus is to the origin, the smaller will this remainder be, so this region of chaotic motion becomes very narrow as the resonant region collapses on the origin.

There remains the all-important question of whether any tori subsist between this dense set of resonances, and indeed whether the islands themselves are not actually swept away by the homoclinic windings of the separatrices. Note that this whole analysis now can be applied to the stable periodic satellite orbit. We then obtain dense resonances among the islands, containing in turn stable periodic orbits and so on. It turns out that typical stable motion is actually harder to analyse than the chaotic unstable motion, our principal subject until now. The preeminent result in this area is the theorem of Kolmogorov, Arnold and Moser, to be discussed in chapter 6, after the preliminary consideration of maps of the circle in chapter 5.

Exercises

1. Verify that the transformation (4.66) also eliminates the time dependence of all the higher-order resonant terms. Therefore the resonant normal form is integrable up to any order.
2. Deduce the normal form for the neighbourhood of a stable equilibrium of an autonomous system with two freedoms.
3. Show that, in the case of an exact resonance in exercise 2, a coordinate transformation will eliminate the dependence of the Hamiltonian on one of the angles. The corresponding action is an independent constant of the motion.

The resonances of order $k \leq 4$ have different structures. The case where $k = 3$ is analysed by Arnold (1978, app. 7). Only a single unstable satellite orbit appears, corresponding to three fixed points. However, satellite fixed points exist on both sides of the resonance (fig. 4.9). The fourth-order resonance can happen generically in two forms. One follows the higher-order resonances, whereas the other is analogous to the third-order resonance.

The resonances of order 1 and 2 are the hardest to analyse. The problem is that it becomes impossible to separate clearly the quadratic part of the Hamiltonian as in (4.34) before starting the analysis. These cases were sorted out by Meyer (1970) in his complete analysis of generic bifurcations of area-preserving maps. Just as with the unstable points studied in section 4.3, there is a unique relation between the normal form for a periodic Hamiltonian and its Poincaré section. The normal forms for all the generic bifurcations of fixed points of area-preserving maps are shown in fig. 4.9.

According to the figure we have second-order resonances for both stable and unstable fixed points. These are known as *period-doubling bifurcations*, since the satellite periodic orbits have twice the period of the central orbit.

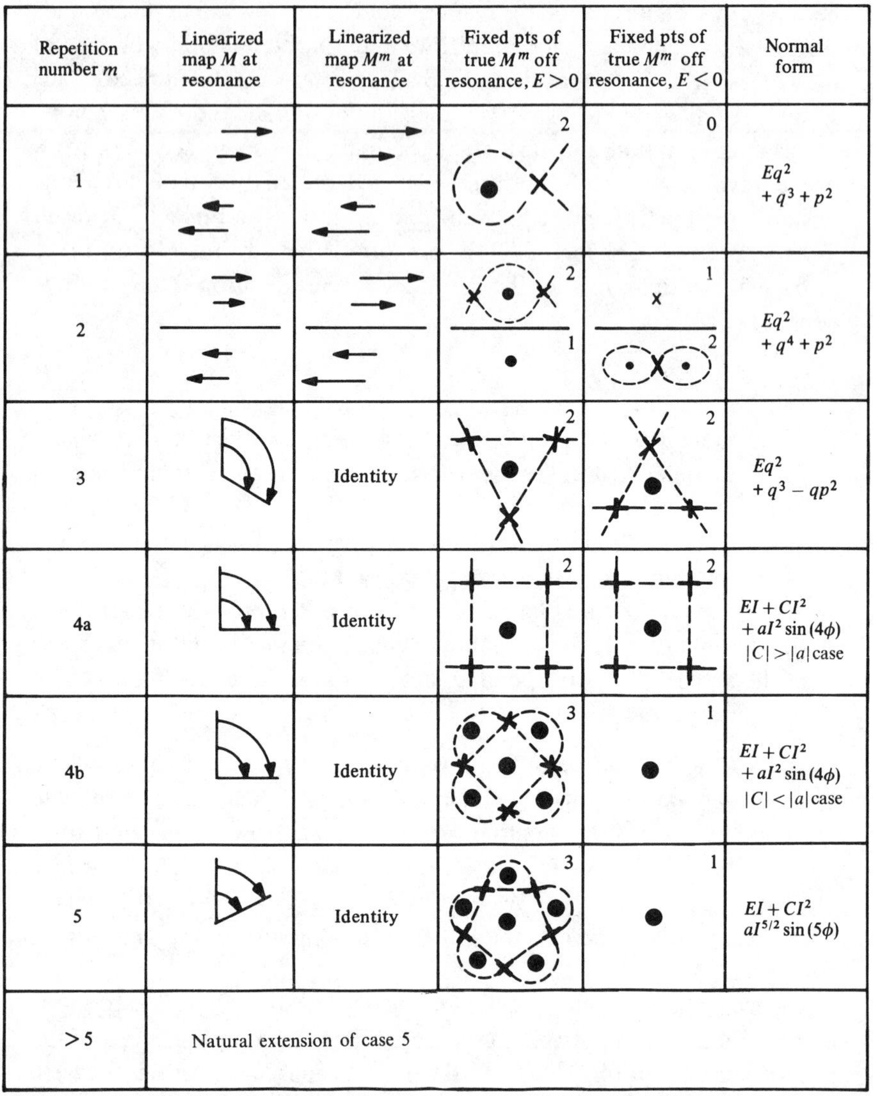

Repetition number m	Linearized map M at resonance	Linearized map M^m at resonance	Fixed pts of true M^m off resonance, $E > 0$	Fixed pts of true M^m off resonance, $E < 0$	Normal form
1			2	0	$Eq^2 + q^3 + p^2$
2			2 1	1 2	$Eq^2 + q^4 + p^2$
3		Identity	2	2	$Eq^2 + q^3 - qp^2$
4a		Identity	2	2	$EI + CI^2 + aI^2 \sin(4\phi)$ $|C| > |a|$ case
4b		Identity	3	1	$EI + CI^2 + aI^2 \sin(4\phi)$ $|C| < |a|$ case
5		Identity	3	1	$EI + CI^2$ $aI^{5/2} \sin(5\phi)$
> 5	Natural extension of case 5				

Fig. 4.9. Normal forms for all the generic bifurcations of periodic orbits. The first column specifies the ratio of the periodicity of the bifurcating orbit to the original orbit. For $n > 5$ all orbits follow the same pattern with pairs of stable and unstable satellite orbits.

In the same way, we can refer to the nth-order resonance as a *period n-ling bifurcation*. It is common for the period doubling of a stable periodic orbit to be followed by a cascade of successive period doublings of the stable periodic orbits, resulting from each bifurcation. This event has been analysed by Bountis (1981), Greene et al. (1981) and Collet, Eckmann and Koch (1981) for area-preserving maps, though its scaling properties were first studied in dissipative maps by Feigenbaum (1978, 1979). In a finite parameter range (energy in Poincaré sections), there occurs an infinite number of bifurcations, resulting in an infinite number of unstable periodic orbits of arbitrarily high period. The previously stable region thus turns into a domain of hyperbolic chaotic motion, as studied in chapter 3. For dissipative maps this is often refered to as the 'universal route to chaos'. We have seen that it is not the only route for conservative maps, where chaotic motion may appear through the homoclinic bifurcation.

The situation for the quadratic map (4.57) is illustrative. For low values of the parameter α, there is an elliptic fixed point, as well as the hyperbolic fixed point at the origin (fig. 4.10). This starts to undergo period-doubling bifurcations for $\alpha > \alpha_0 \simeq 1.67$. However, for this parameter the homoclinic windings of the separatrices have already swept away practically all the invariant curves surrounding the stable point. Thus, in this case period-doubling bifurcations appear only as a final step in the path to completely chaotic motion.

Meyer classified all the generic bifurcations of periodic orbits in area-preserving maps, that is, for all periodic orbits with 'nothing special' about them. Does the classification apply to the reversible maps introduced in section 2.5, or to maps with other symmetries? We found there that an inversion **R**, for which **RR = 1**, is invariant with respect to a coordinate transformation. So if a map has, say, a reflection symmetry, its normal form will also have an inversion symmetry. Furthermore, this symmetry is asymptotic to a reflection for points close to the origin, because the normal-form transformation reduces there to the identity. Therefore, only symmetric terms of the normal form will have nonzero coefficients. Symmetries may therefore cancel the generic resonant term of the normal form. We must then look for the lowest resonant terms that are not cancelled, obtaining in this way a resonant normal form for the symmetric orbit, as in Aguiar, Malta, Baranger and Davies (1987).

Exercises

1. Show that a single reflection symmetry does not alter the normal form for resonances of order $n \geq 3$.

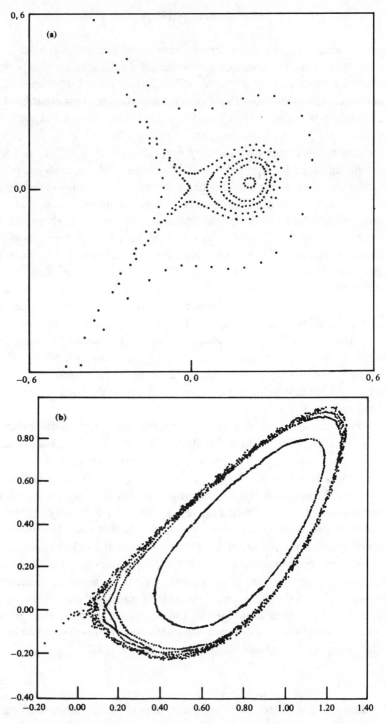

Fig. 4.10. Orbits of the quadratic map (4.57) for $\alpha = 0.2$ (a) and $\alpha = 0.9$ (b).

2. Show that a pair of reflection symmetries cancels the coefficient of the lowest resonant term for $n = 3$. The symmetric resonant normal-form Hamiltonian has level curves resembling the sixth-order resonance, but there are two pairs of satellite orbits of period three.

Whatever the symmetry properties of the map, the possibilities for a bifurcation are limited by a geometric constraint: The *Poincaré index* of a closed curve C that passes no fixed points is defined by Greene et al. (1981) as the number of times that the vector $\mathbf{T}(\mathbf{x}) - \mathbf{x}$ encircles the origin as $\mathbf{x}$ traverses C, taken to be positive or negative, respectively, as the encirclement and traversal are in the same or in opposite directions. For example, the index of a small curve around a single ordinary hyperbolic, elliptic or inversion hyperbolic point is respectively -1, 1 or 1. Provided that C crosses no fixed points, the index is a continuous function of C. Being integer-valued, it must be constant. Thus, we speak of the index of a fixed point, meaning the index of any closed curve surrounding it and no other fixed points. The index is summable – that is, the index of a curve, formed by traversing first one and then another curve, is the sum of the separate indices, so the index of any curve is the sum of the indices of the fixed points that it contains. Continuity of the map in its parameter then implies that the sum of the indices of the fixed points in a given region is constant as the parameter is varied, so long as none of them crosses the boundary of the chosen region. This limits the ways in which bifurcations can take place.

Exercise

Show that in a period n-ling bifurcation, with $n > 2$, elliptic and ordinary hyperbolic orbits must be created in pairs, whatever the symmetry constraints.

5

Maps on the circle

The motion along thin tori surrounding stable periodic orbits in the Birkhoff approximation was studied in section 1.3. The Poincaré maps on such tori are translations of the circle (rotations). According to the analysis in section 4.4, the periodic Birkhoff tori and their neighbours are broken up by the nearly resonant terms of the Hamiltonian, which generate small denominators in the normal-form transformation. Even so, the survival of nonperiodic tori is not excluded, as will be confirmed by the theorem of Kolmogorov, Arnold and Moser in chapter 6. Understanding the motion near stable periodic orbits thus requires a preliminary study of general motion on tori, or their Poincaré sections – maps on the circle.

There is a didactic incentive that perhaps even outweighs the above-mentioned physical motivation for the study of circle maps. In the context of these maps, many of the mathematical difficulties that beset the analysis of stable motion manifest themselves in their clearest, simplest form. We will analyse the effect of the near commensurability of frequencies, that is, rational rotation numbers. The attempt to reduce general maps to rotations leads once again to small denominators. Convergence in their presence is possible, but this result is so surprising that considerable effort will be spent in sketching proof.

5.1 The rotation number

Consider the general dynamical system

$$\dot{\boldsymbol{\theta}} = \boldsymbol{\omega}(\boldsymbol{\theta}), \tag{5.1}$$

defined on the two-dimensional torus $0 \leq \theta_j \leq 2\pi$. The orbits are given by

$$\frac{d\theta_2}{d\theta_1} = \frac{\omega_2(\boldsymbol{\theta})}{\omega_1(\boldsymbol{\theta})} \tag{5.2}$$

if $\omega_1(\boldsymbol{\theta}) \neq 0$. The Poincaré map $T:\theta_2 \rightarrow \theta_2'$ associates to each point θ_2 (and $\theta_1 = 0$) the solution of (5.2) for $\theta_1 = 2\pi$ (fig. 5.1). This diffeomorphism

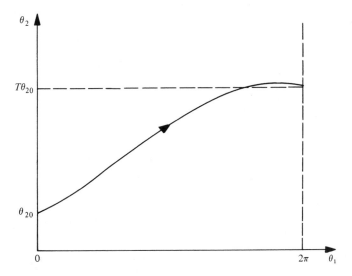

Fig. 5.1. The Poincaré map associates to each point θ_{20} the intersection of its orbit with the line $\theta_1 = 2\pi$.

can always be written in the form

$$T\theta = \theta + F(\theta), \qquad (5.3)$$

dropping the index of θ_2. The angular function $f(\theta)$ satisfies

$$F(\theta + 2\pi) = F(\theta) \qquad \text{and} \qquad dF/d\theta > -1. \qquad (5.4)$$

The rotation number for the map T is defined as

$$\mu = \frac{1}{2\pi} \lim_{k \to \infty} \frac{F(\theta) + F(T\theta) + \cdots + F(T^{k-1}\theta)}{k}. \qquad (5.5)$$

In the case where T is a rotation, $2\pi\mu$ will therefore be just the rotation angle. It is important to show that the above limit exists and that it is independent of θ. The following proof is borrowed from Arnold (1982).

Let us consider the angle by which the point θ is rotated under the kth power of the map T:

$$F_k(\theta) = F(\theta) + F(T\theta) + \cdots + F(T^{k-1}\theta). \qquad (5.6)$$

Since $T(\theta + 2\pi) = T(\theta) + 2\pi$ and hence $T^k(\theta + 2\pi) = T^k(\theta) + 2\pi$, it follows that, if $|\theta_1 - \theta_2| < 2\pi$, then

$$|F_k(\theta_1) - F_k(\theta_2)| < 2\pi. \qquad (5.7)$$

Furthermore, because F_k is 2π-periodic, we can add any multiple of 2π to either θ_1 or θ_2, so that in fact (5.7) holds for any θ_1 and θ_2. Taking now m_k to be the integer for which

$$2\pi m_k \leq F_k(0) \leq 2\pi(m_k + 1), \tag{5.8}$$

we obtain from (5.7)

$$|F_k(\theta) - 2\pi m_k| < 4\pi \tag{5.9}$$

and hence

$$\left|\frac{F_k(\theta)}{2\pi k} - \frac{m_k}{k}\right| < \frac{2}{k}. \tag{5.10}$$

So far we have shown only that $F_k/2\pi k$ is contained in an interval $\sigma_k = ((m_k - 2)/k, \ (m_k + 2)/k)$, which vanishes with k^{-1}. This does not necessarily imply that m_k/k has a constant limit. However, we now note that

$$\frac{F_{kl}(\theta)}{2\pi kl} = \frac{1}{l} \sum_{i=0}^{l-1} \frac{F_k(T^i\theta)}{2\pi k}. \tag{5.11}$$

Each of the terms $F_k(T^i\theta)/2\pi k$ is contained in the interval σ_k, so that

$$\left|\frac{F_{kl}(\theta)}{2\pi kl} - \frac{m_k}{k}\right| < \frac{2}{k}. \tag{5.12}$$

Now $F_{kl}(\theta)/2\pi kl$ also belongs to the interval σ_l, which tends to zero with l^{-1}. It follows that all pairs of intervals σ_k and σ_l intersect one another. Since they both tend to zero, the unique common point of intersection is the rotation number itself.

An immediate corollary is that a map with a periodic orbit must have a rational rotation number. After all, if θ_0 returns after r iterations, its image on a periodic straight line under T^r will be $\theta_0 + 2\pi s$, where s is an integer. Therefore, $F_{lr}(\theta_0) = 2\pi sl$ for all l, so the rotation number is just s/r. Conversely, if $F_r(\theta) > 2\pi s$ for all θ, we cannot have s/r as the rotation number, the same being the case if $F_r(\theta)$ is always smaller than $2\pi s$. Therefore, $\mu = s/r$ implies that $F_r(\theta) = 2\pi s$ for some θ. Thus, a rational rotation number always implies the existence of at least one periodic orbit.

If the rotation number is irrational, the map has no periodic orbit. This means that, for all θ, $F_r(\theta) > 2\pi s$, if $\mu > s/r$, no matter how good an approximation s/r may be to μ. For each θ the order of the points $(\theta, T\theta, \dots, T^k\theta)$ on the circle is the same as for the irrational rotation $2\pi\mu$. Otherwise, the switching of the order of θ and $T^r\theta$ in an interval (θ_1, θ_2) would imply the existence of a periodic orbit in this interval.

5.2 Reduction of circle maps to rotations

We have seen how an irrational rotation number imposes a strong restriction on a differentiable mapping. It is therefore plausible that there exists a coordinate transformation that reduces the map to a mere rotation. This is indeed the content of *Denjoy's theorem* (see Arnold, 1982, or Guckenheimer and Holmes, 1983, for a proof): If a twice-differentiable diffeomorphism of the circle has an irrational rotation number μ, it is topologically equivalent to a rotation of the circle by the angle $2\pi\mu$.

There exists then a continuous transformation capable of reducing the map to a rotation, but this transformation is not necessarily differentiable. Denjoy's theorem is somewhat analogous to the Hartman–Grobman theorem. The latter guarantees a local topological equivalence between a map in the neighbourhood of a hyperbolic point and its linearization, whereas the former guarantees a global equivalence for circle maps. In both cases, a problem arises only if we insist that the transformations be differentiable.

Surprisingly, perhaps, there is another theorem (Arnold, 1982) which asserts that only diffeomorphisms of the circle with rational rotation numbers are structurally stable. A sufficient condition for this is that all the periodic orbits be nondegenerate, that is, that the derivative of the rth powers of the map be nonzero. In other words, if we consider a one-parameter family of maps, a map with nondegenerate periodic orbits is topologically equivalent to all the maps in a given neighbourhood. All of these must have the same rotation number. Conversely the neighbourhood of a map with an irrational rotation number will contain maps that do not coincide with it under any coordinate transformation. Let α be the parameter; then $\mu(\alpha)$ is a function that is constant for some interval of α at every rational value of μ. Since the rational numbers are dense among the real numbers, it is appropriate to wonder whether there are any points α for which $\mu(\alpha)$ is irrational. If the answer is negative, then $\mu(\alpha)$ will have a graph that is a staircase with finite irregular steps. Otherwise, the values of α for which $\mu(\alpha)$ are irrational form a Cantor set. The graph of $\mu(\alpha)$ will then be a *devil's staircase* with infinitely small steps, as described by Mandelbrot (1982).

In this confusing situation it is best to analyse a concrete example (Arnold, 1965):

$$T_\alpha(\theta) = \theta + \alpha + \varepsilon \sin \theta \tag{5.13}$$

in the limit where $\varepsilon \to 0$. We proceed by determining the region in the (α, ε) plane, for which there exists a periodic orbit of period $\mu = s/r$. For $\varepsilon = 0$, there is just the single point $\alpha = 2\pi s/r$. Thus, we seek the small parameters ε and $\alpha' = \alpha - 2\pi s/r$, for which the equation

$$\theta = T^r(\theta) = \theta + 2\pi s + r\alpha' + \varepsilon \sum_{j=0}^{r-1} \sin(T^j\theta) \qquad (5.14)$$

has a solution.

For $\mu = 1$, (5.14) reduces to

$$0 = \alpha' + \varepsilon \sin \theta, \qquad (5.15)$$

wherefore the straight segments

$$|\alpha'| = \varepsilon \qquad (5.16)$$

bound the region with $\mu = 1$ (or 0). For $\mu = \frac{1}{2}$, (5.14) becomes

$$0 = 2\alpha' + \varepsilon \sin \theta + \varepsilon \sin(\theta + \pi + \alpha' + \varepsilon \sin \theta). \qquad (5.17)$$

Expanding the last term around $\theta + \pi$, we obtain

$$(2 - \varepsilon \cos \theta)\alpha' = (\varepsilon^2/2) \sin 2\theta. \qquad (5.18)$$

Thus, as $\varepsilon \to 0$, we have the condition

$$|\alpha'| \le \varepsilon^2/4. \qquad (5.19)$$

Exercise

Show that the equation for an orbit of period 3 reduces to

$$3\alpha' - \sin(2\pi/3)\varepsilon^2 = -\varepsilon^3 \cos(\theta + 4\pi/3)\cos(\theta + 2\pi/3)\sin \theta \qquad (5.20)$$

and hence the condition for a periodic orbit is just

$$|\alpha' - \varepsilon^2/2\sqrt{3}| \le \varepsilon^3/4. \qquad (5.21)$$

The regions of the (α, ε) plane where the rotation number is 1, $\frac{1}{2}$, $\frac{1}{3}$ and $\frac{2}{3}$ are indicated in fig. 5.2. For $r > 1$ they touch the α axis in a cusp, which becomes narrower as r increases. For larger r we still find that the interval of α for which the map has rotation number s/r is $O(\varepsilon^r)$. Since $1 \le s < r$, there are $r - 1$ such intervals; the total length of the regions where μ is rational is of order $\varepsilon + \varepsilon^2 + 2\varepsilon^3 + \cdots$, a convergent series the sum of which vanishes as $\varepsilon \to 0$. Thus, in this case the parameters α, for which the map has an irrational rotation number, form a Cantor set with finite Lebesgue measure. The function $\mu(\alpha)$ is a devil's staircase. Even so, any interval of α will be crossed by the 'tongues' in fig. 5.2, where μ is rational, so no map with an irrational rotation number is structurally stable.

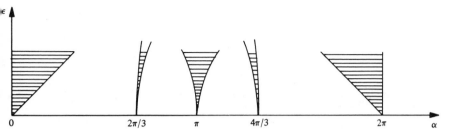

Fig. 5.2. The 'tongues', which reach down to the rational numbers on the x axis, are the parameter ranges with the given rational rotation numbers. The 'tongues' become thinner as the rational denominators increase.

We have seen that the parameter $\alpha/2\pi$ need not coincide with the rotation number, even though the function $f(\theta) = \varepsilon \sin \theta$ satisfies

$$\int_0^{2\pi} f(\theta) \, d\theta = 0. \qquad (5.22)$$

However, for maps with irrational rotation numbers, $\alpha/2\pi$ is indeed the rotation number. In these cases Denjoy's theorem guarantees that the map can be transformed into a rotation for which time averages and configurational averages coincide, so $\bar{f}(\theta) = 0$. The definition (5.5) of the rotation number is just the average of $|\alpha + f(\theta)|/2\pi$, so the rotation number is exactly $\alpha/2\pi$. Conversely, if $\alpha/2\pi$ is an irrational rotation number, then $f(\theta)$ has zero average in a map with the form of (5.13). The picture is then that, for each irrational number times 2π along the line $\varepsilon = 0$ in fig. 5.2, there is a vertical segment along which the map has the same irrational rotation number. The height of this segment will be limited by some spreading 'rational tongue'.

The possibility that T_α has an irrational rotation number depends on α being 'sufficiently irrational' in some sense. Denjoy's theorem then guarantees that a homeomorphism will turn T_α into a mere rotation. In section 5.4 we will find that for sufficiently small ε there may even be analytic reductions of T_α into a rotation. However, we must first examine more closely this concept of being 'sufficiently irrational'.

5.3 Approximating irrational numbers by rational numbers

Irrational numbers can be approximated by rational numbers to arbitrarily high accuracy. For example π is approximated by the sequence

$$\frac{s}{r} = \frac{3}{1}, \frac{31}{10}, \frac{314}{100}, \frac{3142}{1000}, \frac{31416}{10000}, \dots, \qquad (5.23)$$

which provides better approximations as r increases. Thus, for decimal approximations we obtain

$$\left| \mu - \frac{s}{r} \right| < \frac{1}{r}, \tag{5.24}$$

with $r_n = 10^n$, and a similar inequality holds for any other basis.

For a finite sequence of n terms, some bases will lead to better approximations than others, but extending the sequence to higher terms we find that this advantage will be reversed in favour of another basis. Thus, it is worthwhile to look for a basis-independent sequence, such as the sequence of *continued fractions*:

$$\mu_n = a_0 + \cfrac{1}{a_1 + }$$

$$+ \cfrac{1}{a_n}. \tag{5.25}$$

These *approximants*

$$\mu_n = s_n / r_n \tag{5.26}$$

obey the following recursion relation,

$$\begin{aligned} s_n &= a_n s_{n-1} + s_{n-2}, \\ r_n &= a_n r_{n-1} + r_{n-2}, \end{aligned} \tag{5.27}$$

for $n \geq 2$.

This result is demonstrated in Khinchin (1963), a source of many other important results on continued fractions. We proceed by induction, supposing that (5.27) holds up to order $n-1$. Defining

$$\frac{s'_{n-1}}{r'_{n-1}} = a_1 + \cfrac{1}{a_2 + }$$

$$+ \cfrac{1}{a_n}, \tag{5.28}$$

we have

$$s_n = a_0 s'_{n-1} + r'_{n-1},$$
$$r_n = s'_{n-1} \tag{5.29}$$

and by hypothesis

$$s'_{n-1} = a_n s'_{n-2} + s'_{n-3},$$
$$r'_{n-1} = a_n r'_{n-2} + r'_{n-3}. \tag{5.30}$$

Introducing (5.30) into (5.29) and rearranging the terms, we immediately retrieve (5.27), which evidently holds for $n = 2$.

The sequence of integers a_n determines the sequence of approximants (r_n, s_n). Consider the graphical construction, presented in Arnold (1982) and shown in fig. 5.3, by which the pairs (r_n, s_n) are pictured as the integer coordinates of points on the plane. The real numbers are associated with the slope of straight lines passing through the origin. Thus, an irrational number corresponds to a straight line that avoids all the points on the

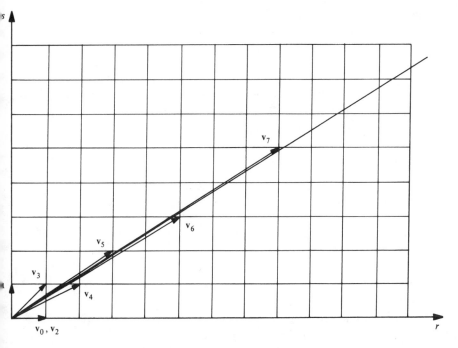

Fig. 5.3. The sequence of vectors v_n supplies ideal approximations to the irrational slope of the straight line.

integer lattice in fig. 5.3. A good sequence of approximants to the irrational number μ is the sequence of vectors $\mathbf{v}_n = (r_n, s_n)$, which starts with $\mathbf{v}_0 = (1, 0)$ and $\mathbf{v}_1 = (0, 1)$. Vectors $\mathbf{v}_n$ are obtained by adding to $\mathbf{v}_{n-2}$ the vector $\mathbf{v}_{n-1}$ multiplied by the largest integer a_n for which $\mathbf{v}_n = a_n \mathbf{v}_{n-1} - \mathbf{v}_n$ still remains on the same side of the straight line as $\mathbf{v}_{n-2}$. The resulting approximant (r_n, s_n) satisfies the recursion relation (5.27); hence, the sequence $a_0, a_1, \ldots$ determines its unique continued-fraction representation. In the limit where $n \to \infty$ we obtain the irrational number as an infinite continued fraction.

How do the continued-fraction approximations compare with those depending on a basis? The difference between two approximants μ_n and μ_{n+1} is

$$\left| \frac{s_n}{r_n} - \frac{s_{n+1}}{r_{n+1}} \right| = \left| \frac{s_n r_{n+1} - s_{n+1} r_n}{r_n r_{n+1}} \right| = \frac{1}{r_n r_{n+1}}, \tag{5.31}$$

since the numerator of the second term is the area of the parallelogram formed by the vectors $\mathbf{v}_n$ and $\mathbf{v}_{n+1}$. This area is 1 for the pair $(\mathbf{v}_0, \mathbf{v}_1)$, and successive parallelograms have a common side and the same height. The approximants are alternately greater and smaller than μ and $r_{n+1} \geq r_n$, so they all satisfy the inequality

$$\left| \mu - \frac{s}{r} \right| < \frac{1}{r^2}. \tag{5.32}$$

Thus, the continued-fraction sequence converges much more rapidly onto an irrational number than any basis-dependent sequence. The larger the numbers a_n, the better the convergence; after all, some $a_n = \infty$ if μ is a rational number. In contrast, the hardest number to approximate by rationals is *the golden mean*,

$$\mu = \cfrac{1}{1 + \cfrac{1}{1 + \cfrac{1}{\cdots}}} = \frac{\sqrt{5} - 1}{2}. \tag{5.33}$$

Can we find a sequence of approximants satisfying even more favourable *Diophantine conditions* than (5.32)? The answer is that (5.32) is the best that can be guaranteed uniformly for all the real numbers. This follows from the result that, for most numbers μ in the interval $[0, 1]$ and for all $\sigma > 0$, there exists a number $K(\mu, \sigma)$, such that

$$\left| \mu - \frac{s}{r} \right| \geq \frac{K}{r^{2+\sigma}} \tag{5.34}$$

for all pairs of integers (r, s). Any infinite sequence of approximants for μ must have $r_n \to \infty$, so that beyond a certain element $K/r_n^{2+\sigma} > 1/r_n^{2+\sigma'}$, whatever the number K. Therefore, since $0 < \sigma' \leq \sigma$ and σ can be made as small as we like, the numbers satisfying the Diophantine conditions (5.34) cannot be approximated to better accuracy than (5.32).

To demonstrate (5.34) we first fix s and r. Consider now the set of numbers μ for which these inequalities are violated. They form an interval of length $2K/r^{2+\sigma}$. The union of all these intervals with the same r has a total length equal to or smaller than $2K/r^{1+\sigma}$, since $s \leq r$. Summing over all the r's we obtain a set, the Lebesgue measure of which is CK, where

$$C = 2 \sum_{r=1}^{\infty} r^{-(1+\sigma)} < \infty. \tag{5.35}$$

By choosing K arbitrarily small, we make the measure of the set of points that violates (5.34) as small as we like. Thus, the set of numbers that infringes these Diophantine conditions for any K has zero measure.

The numbers that satisfy the conditions (5.34) for a given pair of $\sigma > 0$ and $K > 0$ are known as *numbers of type* (K, σ).[1] This is the basis for the vague concept of 'sufficiently irrational' used in the previous sections; that is, the numbers of type (K, σ) with larger maximal K or smaller minimal σ may be thought of as the 'more irrational'.

5.4 Kolmogorov's method

According to Denjoy's theorem, a map with an irrational rotation number is always conjugate to a rotation by a homeomorphism. Just as with the Hartman–Grobman theorem, the proof is nonconstructive, so we now seek the required transformation in a manner analogous to the method of normal forms, that is, by iterated simplifications.

Consider a map of the form

$$\theta \to T\theta = \theta + 2\pi\mu + f(\theta). \tag{5.36}$$

As we saw in section 5.2, the only requirement on $f(\theta)$ for the irrational parameter μ to be the rotation number is that this function have zero

---

[1] Obviously a number of type (K, σ) is also a number of type (K', σ) or (K, σ') if $K' \leq K$ and $\sigma' \geq \sigma$.

average. We will also suppose that $f(\theta)$ is an analytic periodic function
in the strip $|\text{Im } \theta| < \rho$ of the complex plane, where

$$\sup|f(\theta)| \equiv \|f\|_\rho < \varepsilon. \tag{5.37}$$

Exercise

Determine the bound on the function $f(\theta)$ in (5.13) for the strip $|\text{Im } \theta| < \rho$.

We now seek a diffeomorphism

$$\phi \to G\phi = \phi + g(\phi) = \theta, \tag{5.38}$$

which brings the map T to the rotation U (by the angle $2\pi\mu$). Thus, G
satisfies

$$T \cdot G\phi = G \cdot U\phi, \tag{5.39}$$

leading to the functional equation

$$g(\phi + 2\pi\mu) - g(\phi) = f(\phi + g(\phi)). \tag{5.40}$$

The problem is to establish for what rotation numbers (5.40) has a solution.
If f is of order ε, it is plausible that g is too. So expanding f, we obtain
as a first approximation to G the transformation

$$\phi \to G_0\phi = \phi + g^0(\phi) = \theta, \tag{5.41}$$

where the function $g^0(\phi)$ satisfies the equation

$$g^0(\phi + 2\pi\mu) - g^0(\phi) = f(\phi). \tag{5.42}$$

If we expand $f(\phi)$ and $g^0(\phi)$ in Fourier series,

$$f(\phi) = \sum_l f_l \exp(il\phi)$$

and $\tag{5.43}$

$$g^0(\phi) = \sum_l g_l^0 \exp(il\phi),$$

equation (5.42) reduces to

$$\sum_l g_l^0 [\exp(i2\pi l\mu) - 1] \exp(il\phi) = \sum_l f_l \exp(il\phi), \tag{5.44}$$

with the formal solution

$$g_l^0 = \frac{f_l}{\exp(i2\pi l\mu) - 1}. \tag{5.45}$$

There is no exact resonance among the coefficients, because the rotation number was assumed irrational. However, there will be arbitrarily small denominators on the right side, just as with the normal forms of chapter 4. We have now developed the tools with which to tackle such a tricky convergence problem. If μ is of type (K, σ), that is, it satisfies the Diophantine conditions (5.34), the distance of $l\mu$ to the nearest integer is greater than $K/|l|^{1+\sigma}$. The numbers 1 and $\exp(i2\pi l\mu)$ are separated by the arc $2\pi l\mu$ along the unit circle. We seek a lower bound for the length of the chord that joins them. Since the minimum ratio between chord and arc is $2/\pi$, the former satisfies the inequality

$$|\exp(i2\pi l\mu) - 1| \geq \frac{2}{\pi}(2\pi l\mu) \geq \frac{4K}{|l|^{1+\sigma}}. \qquad (5.46)$$

The convergence of the Fourier series for $g^0(\phi)$ depends on the exponential decay of the coefficients f_l being capable of dominating the power law decay of the denominators allowed by (5.46). This can be shown directly from the definition

$$f_l = (2\pi)^{-1} \int_0^{2\pi} f(\phi) \exp(-il\phi) \, d\phi. \qquad (5.47)$$

We can move the path of integration to $\operatorname{Im} \phi = -\rho$ for $l > 0$, since f is analytic in a strip of this width and the vertical integrals along $\operatorname{Re} \phi = 0$ and 2π cancel each other. Thus,

$$f_l = (2\pi)^{-1} \int_0^{2\pi} f(\phi - i\rho) \exp(-il\phi - l\rho) \, d\phi \leq \varepsilon \exp(-|l|\rho), \qquad (5.48)$$

using (5.37). Conversely, for a given sequence of coefficients f_l satisfying (5.48), we can guarantee that the corresponding Fourier series converges to an analytic function $f(\phi)$ in a strip of width 2ρ. We can even determine a bound on $g^0(\phi)$ in the *narrower strip* of width $2(\rho - \delta)$:

$$\|g^0\|_{\rho-\delta} \leq \sum_l (\varepsilon/4K) |l|^{1+\sigma} \exp(-|l|\rho) |\exp(il\phi)|$$

$$\leq (\varepsilon/4K) \sum_l |l|^2 \exp(-|l|\rho) \exp[|l|(\rho - \delta)]$$

$$= \frac{\varepsilon}{4K} \frac{d^2}{d\delta^2} \sum_l \exp(-|l|\delta)$$

$$\leq \frac{2\varepsilon}{4K} \frac{d^2}{d\delta^2} \frac{1}{1 - \exp(-\delta)} \leq \varepsilon D \delta^{-3}, \qquad (5.49)$$

assuming that $\sigma < 1$ and $\delta < 1$. Using more delicate majorations Arnold

(1982) obtains the norm

$$\|g^0\|_{\rho-\delta} \le \varepsilon C \delta^{-2-\sigma}, \tag{5.50}$$

in which the number C [like D in (5.49)] depends only on K and σ.

The transformation G_0 reduces T to a rotation up to order ε^2. In analogy to the theory of normal forms, we can attempt to eliminate the difference between T and a rotation up to order ε^3 and so on. The problem is the difficulty of demonstrating the convergence of this procedure.

A fruitful alternative method was proposed by Kolmogorov (1954): Instead of eliminating terms of higher and higher order, we apply the same procedure as described above to T_1, the result of the transformation G_0 on T. This produces a new transformation G_1 and so on. The convergence is improved, in exact analogy to *Newton's method*, for determining the roots of a function. This consists of linearizing the function $f(x)$ about the point x_0, yielding a first approximation x_1 to the root of $f(x)$ (fig. 5.4). Instead of seeking to improve this approximation by taking the quadratic expansion of $f(x)$ about x_0, we now take a new linear approximation about x_1, and this process is iterated indefinitely. It is readily verified that the nth approximation $f(x_n) = O(\varepsilon^{2n})$, instead of $O(\varepsilon^n)$ obtained for approximations by a power series.

We therefore construct the transformation

$$T_1 = G_0^{-1} \cdot T \cdot G_0; \tag{5.51}$$

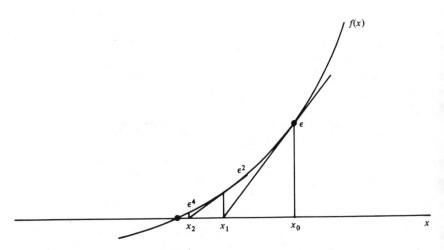

Fig. 5.4. Newton's method for finding the root of a function relies on iterating linear approximations and so converges much better than making polynomial approximations at a fixed initial point.

that is, T_1 is the transformation T viewed with the coordinates ϕ defined by (5.41). If we substituted G for G_0 in (5.51), T_1 would be the rotation U. The problem is now to quantify the statement that T_1,

$$\phi \to T_1\phi \equiv \phi + 2\pi\mu + f_1(\phi), \tag{5.52}$$

is closer to a rotation than T. The *residue* $f_1(\phi)$ between T_1 and U must have zero average just as $f(\phi)$. We estimate $f_1(\phi)$ from the relation $G_0 \cdot T_1 \phi \equiv T \cdot G_0 \phi$:

$$\phi + 2\pi\mu + f_1(\phi) + g^0(\phi + 2\pi\mu + g_1(\phi)) \equiv \phi + g^0(\phi) + 2\pi\mu + f(\phi + g^0(\phi)). \tag{5.53}$$

Hence,

$$f_1(\phi) = [f(\phi + g^0(\phi)) - f(\phi)] - [g^0(\phi + 2\pi\mu + f_1(\phi)) - g^0(\phi + 2\pi\mu)], \tag{5.54}$$

using (5.42) to rearrange the terms. The mean value theorem provides a norm for the first bracket:

$$\| f(\phi + g^0(\phi)) - f(\phi) \|_{\rho - \delta} \leq \| df/d\phi \|_{\rho - \delta} \| g^0 \|_{\rho - \delta}. \tag{5.55}$$

The norm for $df/d\phi$ is easily shown to be $O(\varepsilon\delta^{-2})$ by proceeding as in (5.49). Together with (5.50), we thus obtain

$$\| f(\phi + g^0(\phi)) - f(\phi) \|_{\rho - \delta} \leq C' \varepsilon^2 \delta^{-5}. \tag{5.56}$$

Careful consideration of the second square bracket in (5.54) yields a similar result; that is, we have

$$\| f_1 \|_{\rho - \delta} \leq C'' \varepsilon^2 \delta^{-5}. \tag{5.57}$$

The norm $\| f_1 \|$ estimates the 'error' in the approximation of T_1 by U, a rotation. The order of the 'error' doubles at each successive approximation, just as in Newton's method. Still, it is not yet clear that we can iterate indefinitely these approximations, since a finite choice for δ will 'eat away' the convergence strip ρ after a finite number of iterations, whereas any attempt to reduce δ leads to a small denominator in (5.57). The only way out is to take a diminishing sequence of δ_n's, such as

$$\delta_0 = \delta < \tfrac{1}{2}, \qquad \delta_n = \delta_{n-1}^{3/2}, \qquad \rho_n = \rho_{n-1} - \delta_{n-1}. \tag{5.58}$$

Then

$$\sum_n \delta_n < \delta_0 \sum_n (\delta_0^{1/2})^n < \frac{\rho}{2} \tag{5.59}$$

if δ_0 is sufficiently small.

The final step is to connect the norms $\| f_n \|$ to the sequence of δ_n's. For a sufficiently large integer m, we can replace (5.57) by

$$\| f_1 \|_{\rho - \delta} \leq \varepsilon^2 \delta_0^{-m}. \tag{5.60}$$

But $\delta_n \leq \delta_0$, so the residue of successive approximations satisfies

$$\| f_n \|_{\rho_n} \leq \| f_{n-1} \|^2 \delta_{n-1}^{-m}. \tag{5.61}$$

The number m, common to all the inequalities (5.61), depends only on K and σ. We now specify

$$\varepsilon = \delta_0^N, \tag{5.62}$$

where N is chosen large enough for

$$2N - m \geq 3N/2. \tag{5.63}$$

Thus, using (5.60) and (5.58),

$$\| f_1 \|_{\rho_1} \leq \delta_0^{3N/2} = \delta_1^N, \tag{5.64}$$

whereas repeated use of (5.61) yields

$$\| f_n \|_{\rho_n} \leq \delta_n^N. \tag{5.65}$$

We have shown that Kolmogorov's method converges for $\sigma < 1$ and for small enough $\varepsilon(K)$. Of course, we could have done the same for $\sigma < 2$, obtaining a smaller estimate for ε. The more delicate majorations in the proof of Arnold (1982), on which this exposition is based, provide convergence for a much wider range of $\varepsilon(K, \sigma)$. The range of convergence is actually much wider still, since Herman (1979) proved that even diffeomorphisms not close to rotations can be conjugate to them by a diffeomorphism.

So far it has not been possible to apply Herman's methods to the problems of Hamiltonian dynamics to be studied in the following chapter. The proof of Siegel's theorem also tackles the small-denominator problem, without having to rely on Kolmogorov's method. The latter retains its paramount importance as the basis of the theorem of Kolmogorov, Arnold and Moser. For this reason it has been worthwhile to give a full account of it for the simplest relevant application.

6

Integrable and quasi-integrable systems

In the previous chapter we attained global understanding of a very simple kind of system. We now return to full Hamiltonian systems, but these will be restricted in such a way as to allow a global knowledge of their main characteristics. This class of integrable systems contains all the solved problems in classical mechanics, as well as the truncations of the Birkhoff normal forms studied in chapter 4. We will find that no limitation need be made on the number of freedoms of an integrable system. In fact, their definition can be extended up to an infinite number of freedoms, leading to the solution of important partial differential equations, but this subject lies beyond the scope of this book.

We start by defining integrable systems and by studying the geometry of their invariant surfaces. This leads to the definition of general action-angle variables. The consideration of a few simple examples elicits the concept of caustics, that is, singularities of the projections of invariant surfaces. We will discuss briefly Thom and Arnold's classification of some of the simpler generic caustics.

In conclusion we will study perturbations of integrable systems. In section 6.5 we will discuss the averaging principle, which is then related to a stationary perturbation theory, reminiscent of the resonant normal forms encountered in section 4.4. This evokes the question of survival of the invariant surfaces of an integrable system after a perturbation, to be answered by the Kolmogorov, Arnold and Moser (KAM theorem). The sketch of its proof relies on that of the analytical reduction of circle maps to rotations presented in section 5.4.

6.1 Invariant tori

A necessary condition for a system with L freedoms to be integrable is that it have L independent constants of the motion. The Hamiltonian of an autonomous system $H(\mathbf{x})$ is a constant function; therefore, there must be

$L-1$ other functions with the property that

$$\frac{d}{dt}F(\mathbf{x}(t)) = \sum_{l=1}^{L}\left(\frac{\partial F}{\partial q_l}\frac{\partial H}{\partial p_l} - \frac{\partial F}{\partial p_l}\frac{\partial H}{\partial q_l}\right) = \{F,H\} = 0, \qquad (6.1)$$

where the penultimate term defines the *Poisson bracket* of the functions F and H. The antisymmetry of the Poisson bracket implies that H will also be a constant of the motion for the Hamiltonian F. A fundamental property (demonstrated in Arnold, 1978, sec. 39) is that the flows $\mathbf{x}_H(t)$ and $\mathbf{x}_F(t')$ generated by two Hamiltonians that satisfy (6.1) commute: Whichever the order in which we 'switch on' the Hamiltonian F for the time t' and H for the time t, the system arrives at the same point $\mathbf{x}$, if it started out from the same point $\mathbf{x}_0$.[1]

The L constants of the motion must be independent; that is, at each point $\mathbf{x}$ the vector fields $\dot{\mathbf{x}}_i = \mathscr{J}\partial F_i/\partial\mathbf{x}$ and $\dot{\mathbf{x}}_j = \mathscr{J}\partial F_j/\partial\mathbf{x}$ [with $\mathscr{J}$ given by (1.37)] must be linearly independent. The last condition defining an *integrable system* is that each pair of functions F_i and F_j be in *involution*:

$$\{F_i, F_j\} = 0. \qquad (6.2)$$

This condition encompasses (6.1) if we identify $H(\mathbf{x}) = F_L(\mathbf{x})$.

Each orbit of an integrable system is contained in an L-dimensional surface T, resulting from the multiple intersection of the L surfaces $F_j(\mathbf{x}) = \text{const}$, each with $(2L-1)$ dimensions. The condition of involution guarantees that we would obtain the same invariant surface by choosing any one of the functions $F_j(\mathbf{x})$ as the Hamiltonian, or even by choosing any arbitrary function of the F_j's. The L Hamiltonians define L vector fields tangent to T, because each vector is tangent to an orbit contained in this surface.

Let us consider again equation (6.1). Comparison with (1.38) establishes that the Poisson bracket of two functions F and H is merely the skew product $\dot{\mathbf{x}}_F \wedge \dot{\mathbf{x}}_H$, that is, the symplectic area or (reduced) action of the parallelogram formed by these two vectors. The L vectors $\dot{\mathbf{x}}_l$, corresponding to the L functions F_l, can be taken as a basis for the tangent plane to the invariant surface T. Hence, the condition of involution implies that the symplectic area is null for any pair of vectors tangent to T. This differential result can evidently be extended to any reducible circuit on T, a property that defines a *null or Lagrangian surface*.

---

[1] If F and H are considered the classical limit of quantum variables $\hat{F}$ and $\hat{H}$ (operators), then (6.1) is a consequence of $[\hat{F},\hat{H}] = 0$. Consequently, the evolution operators $\exp(-i\hat{F}t')$ and $\exp(-i\hat{H}t)$ also commute. The commutation of the classical flow thus follows from the commutation of the quantum flow.

Consider now the orbits defined by the L independent vector fields $\dot{\mathbf{x}}_l$, starting from an arbitrary point $\mathbf{x}_0$ on the surface T. The positions along each orbit result from the action of the one-parameter groups of canonical transformations

$$G_l^{t_l}\mathbf{x}_0 = \mathbf{x}(t_l, \mathbf{x}_0) \tag{6.3}$$

with the properties

$$G_l^{t_l}G_l^{s_l} = G_l^{t_l + s_l} \quad \text{and} \quad (G^{t_l})^{-1} = G^{-t_l}. \tag{6.4}$$

Since these L groups commute,

$$G_k^{t_k}G_l^{t_l} = G_l^{t_l}G_k^{t_k}, \tag{6.5}$$

we can define the action of the commutative group

$$\mathbf{G}^t = G_1^{t_1}\cdots G_L^{t_L}, \tag{6.6}$$

with $\mathbf{t} = (t_1,\ldots,t_L)$.

It is easy to define an invariant measure on T for the motion generated by any of the Hamiltonians F. Each of these preserves the Liouville measure $\delta(F_l(\mathbf{x}) - E_l)\,d\mathbf{x}$; therefore,

$$d\mu = \prod_{l=1}^{L} \delta(F_l(\mathbf{x}) - E_l)\,d\mathbf{x}, \tag{6.7}$$

will be invariant on T. We can thus apply the Poincaré recurrence theorem (section 3.4) to the case of bounded motion, that is, when T is a compact surface. The implication is that, in any neighbourhood of a point $\mathbf{x}_0$, some point $\mathbf{x}_0'$ will return to this neighbourhood after some time t_l. For a group of flows that commute, this result can be strengthened: We can bring $\mathbf{x}_0$ to $\mathbf{x}_0'$ by means of the action of the element $\mathbf{G}^{\delta t}$ of the group $\mathbf{G}^t$. Thus,

$$G_l^{t_l}\mathbf{x}_0 = \mathbf{G}^{-\delta t}G_l^{t_l}\mathbf{G}^{\delta t}\mathbf{x}_0 \tag{6.8}$$

is a point that has returned close to $\mathbf{x}_0$ after the time t_l, under the flow generated by F_l.

We can take the t_i's as coordinates on T by fixing an initial point $\mathbf{x}_0$, because $\mathbf{G}^t$ is a commutative group. This arrangement is sketched in fig. 6.1 for $L = 2$ within a three-dimensional subspace. Consider the orbit γ_1 of $\mathbf{x}_0$ under the action of F_1. It returns to $\mathbf{x}_0'$ in a neighbourhood of $\mathbf{x}_0$ after a time t_1'. In fact, γ_1 must intersect the $t_1 = 0$ coordinate plane in a time t_1^*. To see this, note that γ_1 returns close to $\mathbf{x}_0$, where its velocity $\dot{\mathbf{x}}_1$ is transverse to this plane. The configuration is sketched in fig. 6.2a for $L = 2$. If we now switch off F_1 and switch on any one of the other Hamiltonians F, the system

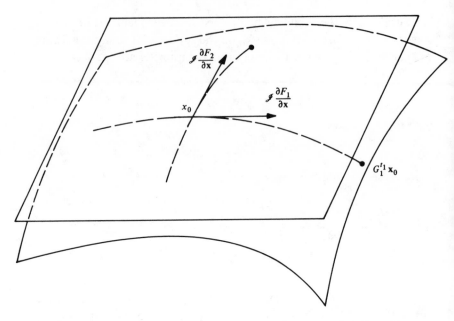

Fig. 6.1. The velocity vectors $DG^t\mathbf{x}_0$ are tangent to the torus and the lines $G_l^{t_1}\mathbf{x}_0$ can be used as coordinates on the torus.

will move on a $\mathbf{t}$ surface which is a new image of $t_1 = 0$. But this motion, without any participation of F_1, determines a plane in the $\mathbf{t}$ coordinates, which reduces to a vertical straight line for $L = 2$ (fig. 6.2b). The plane $t_1 = t_1^*$ contains a new image of the $\mathbf{t}$ origin, $\mathbf{x}_0$. The vector joining the origin to this point generates an irreducible circuit on T, since reducible curves are represented by closed loops in the $\mathbf{t}$ coordinates.

By repeating the same procedure for γ_2, the curve $G_2^{t_2}\mathbf{x}_0$, we obtain the image of the coordinate plane $t_2 = 0$ as a parallel plane with the constant coordinate t_2^* and so on for all the other coordinates. The set of all these planes with $t_j = \pm t_j^*$ forms an L cube centred on the origin. On each of the L pairs of faces there will be a different image of the origin (Fig. 6.2b), such that the vector joining it to the origin corresponds to an irreducible circuit on T. The vectors with tips on $t = \pm t_j^*$ correspond to the same circuit

Fig. 6.2. (a) The coordinate line γ_1 eventually returns close to $\mathbf{x}_0$, with its velocity $\dot{\mathbf{x}}_1$ nearly parallel to $\dot{\mathbf{x}}_1(\mathbf{x}_0)$. (b) In the t coordinates the t_2 axis has multiple images separated by a period t_1^*. Traversals from the origin to its images represent irreducible circuits on the torus T.

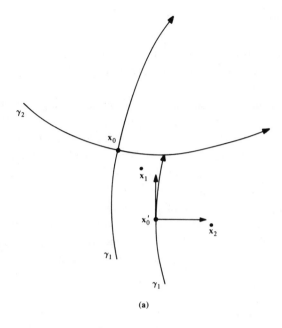

(a)

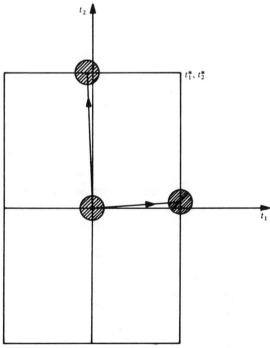

(b)

traversed in opposite senses, and all the linear combinations of these vectors supply multiple windings of the irreducible circuits.

If the neighbourhood surrounding the origin is chosen to be small, then each image of the origin will be close to the centre of the cube face, as shown in fig. 6.2b. However, γ_j may intersect the $t_j = 0$ plane many times just outside this neighbourhood for $t_j < t_j^*$. By increasing the size of the neighbourhood we incorporate these intersections, so that the time t_j^* for the first crossing in the neighbourhood is reduced. The structure is still the same, with the $2L$ images of the origin lying in the faces of a cube (farther from their centre). If the neighbourhood is sufficiently large (for instance, if it covers uniquely the entire surface T), we obtain minimum times t_j^* corresponding to the first traversal of the planes $t_j = 0$.

The advantage is that we now obtain the discrete group of all the irreducible circuits on T. If we label the images of the origin on the faces of the cube by $\mathbf{e}_l$, then the members of the group are specified by

$$\mathbf{t} = n_1 \mathbf{e}_1 + \cdots + n_L \mathbf{e}_L, \tag{6.9}$$

where n_l is an integer. There can be no other points in this group, because there are none inside the cube by construction and we can bring any point outside to the inside by adding to it one of the vectors $\mathbf{t}$ in (6.9). Finally we note that this group is independent of the point $\mathbf{x}_0$, since for any point $\mathbf{x} = \mathbf{G}^r \mathbf{x}_0$,

$$\mathbf{G}^t \mathbf{x} = \mathbf{G}^{t+r} \mathbf{x}_0 = \mathbf{G}^r \mathbf{x}_0 = \mathbf{x}, \tag{6.10}$$

for the $\mathbf{t}$'s satisfying (6.9). We have thus shown that there are exactly L independent circuits passing through any point of the invariant surface. It follows that T is an L torus, that is, a common torus in the case where $L = 2$.

We can now make a linear transformation from the $\mathbf{t}$ coordinates to the $\boldsymbol{\phi}$ coordinates, the axes of which are the vectors $\mathbf{e}_l$, so that ϕ_l varies in the interval $(0, 2\pi)$ from the origin to the tip of the vector. In these coordinates T is the Cartesian product of L circles and the movement generated by the Hamiltonian $H(\mathbf{x})$ is rectilinear and uniform:

$$\boldsymbol{\phi}(t) = \boldsymbol{\phi}(0) + \boldsymbol{\omega} t. \tag{6.11}$$

We conclude that the motion of an integrable system can be reduced to a translation on a torus, the properties of which were studied in section 1.3. This result constitutes the main content of the *Liouville–Arnold theorem*. We will conclude its discussion by showing that a canonical transformation can always be found such that the motion on the torus is given by (6.11).

The obvious choice would be to take the L variables $\mathbf{F}$ and L angles $\boldsymbol{\phi}$ as coordinates for an integrable system. The problem is that the transformation $(\mathbf{p}, \mathbf{q}) \to (\mathbf{F}, \boldsymbol{\phi})$ would not be canonical in general, entailing the loss of the Hamiltonian form of the system. So we must find new constants of the motion $\mathbf{I}(\mathbf{F})$ that are canonically conjugate to the angles $\boldsymbol{\phi}$.

The condition for the transformation $(\mathbf{p}, \mathbf{q}) \to (\mathbf{I}, \boldsymbol{\phi})$ to be canonical is that it preserve the symplectic area for any closed loop in phase space:

$$\oint \mathbf{p} \cdot d\mathbf{q} = \oint \mathbf{I} \cdot d\boldsymbol{\phi}. \tag{6.12}$$

Since the angles $\boldsymbol{\phi}$ have no dimension, it follows that the new variables $\mathbf{I}$ must have the dimension of an action. Each L-dimensional torus is characterized by L irreducible circuits. We can therefore associate the L variables $\mathbf{I}$ to the symplectic area of these circuits, as long as all the possible deformations of a given irreducible loop have the same action. This condition is indeed satisfied by the invariant tori of an integrable system, because all the deformations of an irreducible circuit are obtained by adding to it a reducible loop of zero action. We thus define the *action variables*

$$I_j = (2\pi)^{-1} \oint_{\gamma_j} \mathbf{p} \cdot d\mathbf{q}, \tag{6.13}$$

where γ_j is the jth irreducible circuit on the torus. Provided that the Jacobian determinant $\partial \mathbf{I}/\partial \mathbf{F}$ is nonzero, we can substitute $H(\mathbf{F}) = H(\mathbf{I})$.

It remains to show that the variables conjugate to $\mathbf{I}$ are the required angles $\boldsymbol{\phi}$. To this end we note that the Lagrangian property of the torus implies the existence of the function

$$S(\mathbf{q}, \mathbf{I}) = \int_{\mathbf{q}_0}^{\mathbf{q}} \mathbf{p}_{\mathbf{I}}(\mathbf{q}) \cdot d\mathbf{q}. \tag{6.14}$$

In a two-dimensional phase space the torus specified by a given $\mathbf{I}$ is a closed curve, so that $p_I(q)$ is an even number of points. Generally the L-dimensional torus also intersects transversely the L-dimensional plane $\mathbf{q} = $ const in discrete points within the $2L$-dimensional phase space. The function S is therefore multivalued. If we choose $S(\mathbf{q}, \mathbf{I})$ as the generating function for a canonical transformation $(\mathbf{p}, \mathbf{q}) \to (\mathbf{I}, \boldsymbol{\phi}')$, the transformation is implicitly given by

$$\partial S / \partial \mathbf{q} = \mathbf{p}, \qquad \partial S / \partial \mathbf{I} = \boldsymbol{\phi}'. \tag{6.15}$$

For an irreducible circuit of the torus, γ_i, the change of the angles $\boldsymbol{\phi}'_j$ is simply

$$(\Delta\phi'_j)_{\gamma_i} = \Delta_{\gamma_i}\frac{\partial S}{\partial I_j} = \frac{\partial}{\partial I_j}\Delta_{\gamma_i}S = \frac{\partial}{\partial I_j}2\pi I_j = 2\pi\delta_{ij}. \qquad (6.16)$$

So the conjugate angles $\phi'_j = \phi_j$, the privileged angular coordinates on the torus.

The solution of an integrable system can thus be found in principle by determining the canonical transformation $(\mathbf{p}, \mathbf{q}) \to (\mathbf{I}, \boldsymbol{\phi})$, which is guaranteed to exist. There is no general method for achieving this end, but once the transformation has been found, we obtain Hamilton's equations in the form

$$\dot{\mathbf{I}} = -\frac{\partial H}{\partial \boldsymbol{\phi}}(\mathbf{I}) = 0, \qquad \dot{\boldsymbol{\phi}} = \frac{\partial H}{\partial \mathbf{I}} = \boldsymbol{\omega}. \qquad (6.17)$$

So the motion is given by (6.11), with the frequencies $\boldsymbol{\omega}$ specified by (6.17). Generally the motion will be multiply periodic, admitting the Fourier expansion

$$\mathbf{q}(t) = \sum_{\mathbf{m}} \mathbf{q_m}(\mathbf{I})\exp[i\mathbf{m}\cdot(\boldsymbol{\omega}t + \boldsymbol{\phi}_0)],$$

$$\mathbf{p}(t) = \sum_{\mathbf{m}} \mathbf{p_m}(\mathbf{I})\exp[i\mathbf{m}\cdot(\boldsymbol{\omega}t + \boldsymbol{\phi}_0)]. \qquad (6.18)$$

6.2 Examples of integrable systems

The simplest case is that of a system with a single freedom and a Hamiltonian of the form

$$H = p^2/2m + V(q) \qquad (6.19)$$

for energies $E < V(\infty)$ and $E < V(-\infty)$. The tori are then p-symmetric level curves of the Hamiltonian. The corresponding action variable is just

$$I = (2\pi)^{-1}\oint p\,dq = \pi^{-1}\int_{q_1}^{q_2} dq\{2m[E - V(q)]\}^{1/2}, \qquad (6.20)$$

where the q_j are the *turning points*, for which $p = 0$.

Example. The square well; a ball rolling freely with speed v between two walls separated by a distance d. It is assumed that the ball has elastic collisions with the walls. This 'square well' is more familiar in quantum mechanics. The action is obtained from fig. 6.3 as

$$I = (2\pi)^{-1}(\text{area of rectangle}) = mvd/\pi. \qquad (6.21)$$

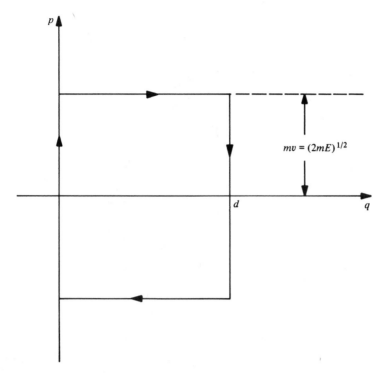

Fig. 6.3. The action variable for the square well is obtained from the area of the rectangle in phase space.

Since the corresponding energy is $E = mv^2/2$, the Hamiltonian is just

$$H(I) = (\pi I)^2/(2md^2). \tag{6.22}$$

In this case the angle variable depends linearly on q, which already has a (piecewise) uniform motion.

Exercise

Show that the dependence of the Hamiltonian $p^2/2m + C\tan^2(\pi/q)$ on the action variable tends to that of (6.22) in the limit of high energies.

Example. The harmonic oscillator. The Hamiltonian has the form

$$H(p, q) = p^2/2 + \omega^2 q^2/2 = E. \tag{6.23}$$

The level curve is an ellipse with semiaxes $(2E)^{1/2}$ and $(2E)^{1/2}/\omega$, so that

$$I = (2\pi)^{-1}(\text{area of an ellipse}) = E/\omega, \tag{6.24}$$

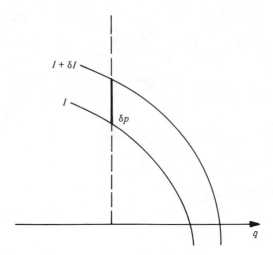

Fig. 6.4. The ratio $\partial\phi/\partial q$ is proportional to the δp separating level curves whose actions differ by δI.

that is,

$$H(I) = \omega I. \tag{6.25}$$

The angular frequency $\dot\phi = \omega$ is therefore the same for all ellipses.

Of course, this is just the case of linear motion around a stable equilibrium already analysed in chapter 1. Through the canonical rescaling $(p, q) \rightarrow (p', q') = (\omega^{-1/2}p, \omega^{1/2}q)$, the Hamiltonian reduces to the standard form $\omega(p'^2 + q'^2)/2$. We hence recognize the action-angle variables as the canonical polar coordinates in the (p', q') plane, defined by (1.49).

The angular variable for a one-freedom system is

$$\phi(q) = \frac{\partial}{\partial I} \int_{q_0}^{q} dq \, p(q, I) \tag{6.26}$$

within an additive constant. We can therefore visualize $\partial\phi/\partial q$ to be proportional to the δp separating level curves that have actions differing by δI (fig. 6.4). Evidently, $\partial\phi/\partial q \rightarrow \infty$ when $q \rightarrow q_j$, a turning point. The angular distribution along the invariant curve can also be understood directly from Liouville's theorem. The conservation of area by the flow implies that the linearly moving angle becomes stretched where invariant curves approach each other and compressed where they separate.

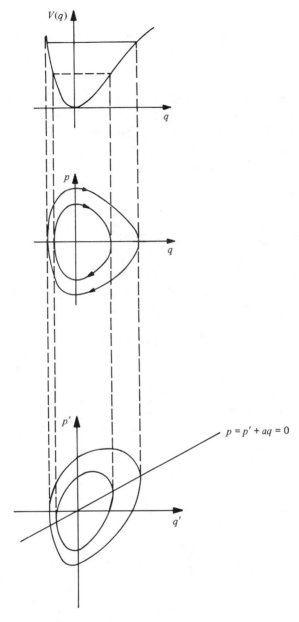

Fig. 6.5. The effect of a canonical shear transformation on the level curves is that the turning points no longer have **p** = 0.

Exercise

Show that, if s is the length along the elliptic level curve of the harmonic oscillator (6.23), then the ratio between $d\phi/ds$ at the tips of the major and minor axes is ω^2.

Other examples and properties of one-freedom systems are discussed in Percival and Richards (1982). Note that general coordinate transformations will destroy the simple form of Hamiltonians (6.19). The turning points will then no longer coincide with the zeros of p along the level curve. One such class of transformations consists of the symplectic (linear) transformations studied in section 1.2. Another one includes the shears

$$p' = p - A(q), \qquad q' = q.$$

The effect of a linear shear on the level curves is shown in fig. 6.5. Evidently, the composition of two canonical transformations is again canonical. In the case of polynomial transformations, Engel (1958) proved that they can always be reduced to the composition of a symplectic transformation with a shear.

Example. Motion under a central force. The energy has the form

$$H(p_r, p_\theta, r, \theta) = \frac{1}{2m}p_r^2 + \frac{1}{2mr^2}p_\theta^2 + V(r) \tag{6.27}$$

for the geometry indicated in fig. 6.6a. The *angular momentum* p_θ is a constant of the motion, because the Hamiltonian is independent of θ. It is readily verified that H and p_θ are in involution, so this two-freedom system is integrable.

Two irreducible circuits for the torus, labeled by a given energy and angular momentum, are the rotation in θ with constant r and the libration in r with constant θ. The corresponding action variables are

$$I_2 = (2\pi)^{-1} \int_0^{2\pi} p_\theta \, d\theta = p_\theta \tag{6.28}$$

and

$$I_1 = \frac{1}{2\pi}\oint p_r \, dr = \frac{1}{\pi}\int_{r_1}^{r_2}\left\{2m\left[E - \frac{1}{2mr^2}I_2^2 - V(r)\right]\right\}^{1/2} dr, \tag{6.29}$$

where r_1 and r_2 are the turning points for the libration shown in fig. 6.6b. The action I_2 may be positive or negative and $H(I_1, I_2)$, obtained from the inversion of (6.29), is symmetric in this variable (fig. 6.7).

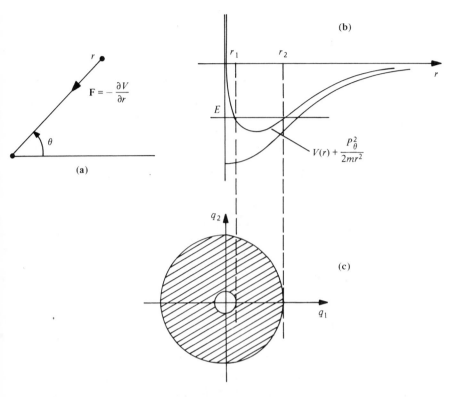

Fig. 6.6. (a) The central force always points in the direction of the origin, and its magnitude depends only on the coordinate r. (b) The turning points for the librations are the same as for the one-dimensional motion with the effective potential $V(r) + p_\theta^2/2mr^2$. (c) The invariant torus projects as an annulus in configuration space; its boundaries are the caustics.

The invariant torus projects into *configuration space* (r, θ) or (q_1, q_2) as an annulus (fig. 6.6c). Each point within it is the image of two points of the torus, with the two values $\pm p_r$ allowed by (6.27) for fixed $H = E$ and p_θ. The boundary of the torus projection is known as the *caustic*. It is composed of the two circles $r = r_1$ and $r = r_2$, where the two pre-images on the torus coalesce. The torus defined by the same energy, but angular momentum $-p_\theta$, has the same caustics.

The Hamiltonian depends nonlinearly on I_1 and I_2, entailing the continuous variation of the frequencies

$$\boldsymbol{\omega} = \frac{\partial H}{\partial \mathbf{I}}(\mathbf{I}) \tag{6.30}$$

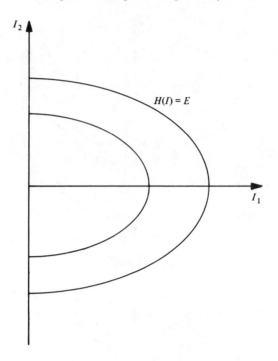

Fig. 6.7. The level curves of $H(I_1, I_2)$ are symmetric in I_2.

among the invariant tori. In any neighbourhood $(\delta I_1, \delta I_2)$ of the action variables we find tori with rationally related frequencies, that is, tori where all the orbits are periodic. The case of the Kepler problem, in which all tori have periodic orbits, is singular – there are other constants of the motion beyond angular momentum and energy. Another instance, in which all the orbits are periodic, is that of the isotropic harmonic oscillator with $V(r) = \omega^2 r^2$. The extra constants of the motion in this case will be discussed in the following example.

Example. Separable systems. These are characterized by Hamiltonians of the form

$$H(\mathbf{p}, \mathbf{q}) = \sum_{l=1}^{L} H_l(p_l, q_l). \qquad (6.31)$$

All the *partial Hamiltonians* H_l are in involution with one another and with the full Hamiltonian H. Choosing the corresponding *partial energies* E_l as constants of the motion, we obtain the action variables I_l independently from equations like (6.20). The most powerful general method for integrating the equations of motion, that of Hamilton and Jacobi, relies on transformations to coordinates that separate the Hamiltonian. This is discussed by Arnold (1978) and Goldstein (1980). It follows that the Hamilton–Jacobi method can be applied only to integrable systems.

In the case of a two-dimensional harmonic oscillator or of a pair of uncoupled oscillators, the Hamiltonian will be

$$H(\mathbf{I}) = \omega_1 I_1 + \omega_2 I_2. \tag{6.32}$$

Its level curves are straight-line segments (fig. 6.8). The frequencies are the same for all tori, so that all orbits are periodic if the ω_j are rationally related. Otherwise, the orbits are all open. In the case where $\omega_1 = \omega_2$, the oscillator is isotropic. The fact that all orbits are then periodic may be attributed to the existence of a further constant of the motion p_θ, independent of H_1 and H_2. The intersection of the level curves of these three functions is a closed curve, which defines the orbit.

The caustics for the torus of a separable system of two freedoms are easily found from the construction shown in fig. 6.9. The independence of the motion in the conjugate planes (p_1, q_1) and (p_2, q_2) implies that the caustics are straight-line segments corresponding to the turning points. Therefore,

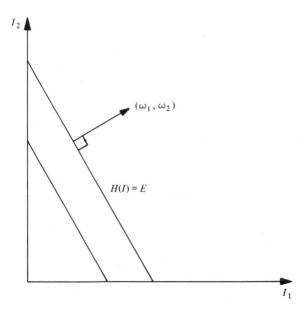

Fig. 6.8. The Hamiltonian for a two-dimensional oscillator is a linear function of the actions.

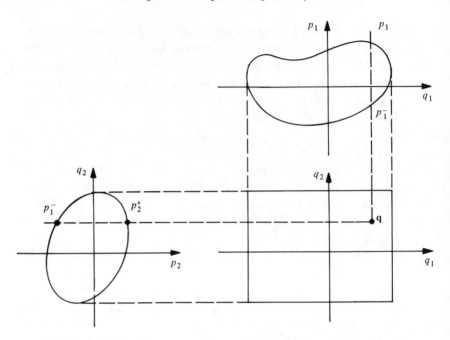

Fig. 6.9. From the phase graphs of the two independent motions of a separable system, we obtain the projection of the torus onto the configuration space. Each point inside the rectangle is the image of four points inside the torus.

the torus projects onto the configuration plane as a rectangle. Any point $\mathbf{q}$ inside it corresponds to four points on the torus with coordinates $(p_1^\pm, q_1^\pm)$. Two pairs coalesce on the (double) caustic. This projection of a smooth torus would be impossible were it embedded in three-dimensional space rather than in four dimensions. We can still 'visualize' this projection by considering a vertical tyre tube (i.e., the position in which it rolls on the ground). When emptied of air, it collapses into a rectangle on the floor, just as our separable torus, but it is no longer smooth. In a four-dimensional space we can now 'fill it out along the fourth dimension' without altering the projection.

A final important example of integrable systems is provided by the normal-form Hamiltonians, which approximately describe the motion surrounding a stable periodic orbit, studied in section 4.4. Truncating the Birkhoff normal form, we obtain an integrable system, the action-angle variables of which are I and J. Even the resonant normal forms are integrable systems and almost all points lie on tori, though this is not true for the points lying along separatrices in the (I, θ) plane. The reason is that these are level curves of the reduced Hamiltonian $H(I, \theta)$ (a constant of the motion) with the same partial energy as one of its equilibria. At such a point the vector field $(\dot{I}, \dot{\theta}) = 0$, contrary to the assumptions in the Arnold–Liouville theorem. The phase space of a system with separatrices cannot be globally transformed to action-angle variables. This is possible only for each separate family of tori.

6.3 Caustics

The previous examples have made familiar the fact that $\mathbf{p}(\mathbf{q})$ for an invariant torus and hence the action $S(\mathbf{q})$, defined by (6.14), are multivalued functions. The caustics limiting the motion in configuration space are the boundaries of the branches of $S(\mathbf{q})$. They correspond to the set of points in phase space where

$$\det d\mathbf{p}/d\mathbf{q} = \infty. \tag{6.33}$$

The way that different branches of $S(\mathbf{q})$ join together at a caustic can be analysed by means of its Legendre transform,

$$S(\mathbf{p}) = \mathbf{p} \cdot \mathbf{q} - S(\mathbf{q}), \tag{6.34}$$

where $\mathbf{q}(\mathbf{p})$ is given by (6.15). In the case of one freedom, we can make the graphical construction shown in fig. 6.10. The Lagrangian condition on the

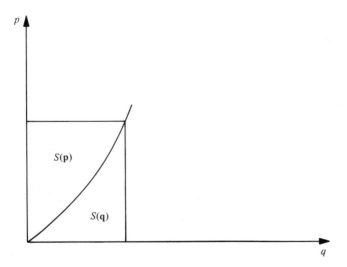

Fig. 6.10. Graphical construction of the action $S(\mathbf{p})$ given $S(\mathbf{q})$; together they make up the area of a rectangle.

invariant torus implies that

$$\oint \mathbf{q} \cdot d\mathbf{p} = -\oint \mathbf{p} \cdot d\mathbf{q} = 0. \tag{6.35}$$

Therefore, the function

$$S(\mathbf{p}) = \int_{\mathbf{p}_0}^{\mathbf{p}} \mathbf{q} \cdot d\mathbf{p} \tag{6.36}$$

is the same function defined by (6.34) within an additive constant. This function is also multivalued, and its branches join at caustics of the projection of the torus onto the **p** space (fig. 6.11). However, $S(\mathbf{p})$ is a one-to-one function in the neighbourhood of the caustics of $S(\mathbf{q})$. Differentiating (6.36),

$$\mathbf{q} = \partial S(\mathbf{p})/\partial \mathbf{p}, \tag{6.37}$$

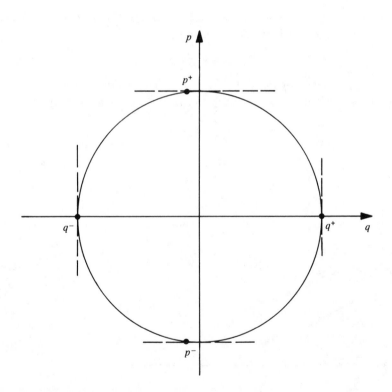

Fig. 6.11. The function $S(\mathbf{p})$ is multivalued like $S(\mathbf{q})$. The branches meet at the caustics p^+ and p^-.

and inverting (6.33), we obtain the caustic condition in the form

$$\det \left| \frac{\partial \mathbf{q}}{\partial \mathbf{p}} \right| = \det \left| \frac{\partial^2 S}{\partial \mathbf{p} \partial \mathbf{p}} \right| = 0. \tag{6.38}$$

The map $\mathbf{p} \rightarrow \mathbf{q}$ is given by the gradient of the function $S(\mathbf{p})$ according to (6.37). The classification of the locally generic forms of the simplest singularities (6.38) of gradient maps is the content of *Thom's theorem*, the cornerstone of *catastrophe theory*, presented in, for example, Poston and Stewart (1978). Generic singularities are those whose form remains invariant with respect to small alterations of the function $S(\mathbf{p})$. The classification depends only on the dimension of configuration space. The simplest case is that of one freedom, for which the only structurally stable caustic is the *fold*, that is, the turning point already met in several examples. Folds also appear in higher-dimensional configuration spaces; this is the case of the fold lines in fig. 6.6c. The dimension of the locus of fold points is

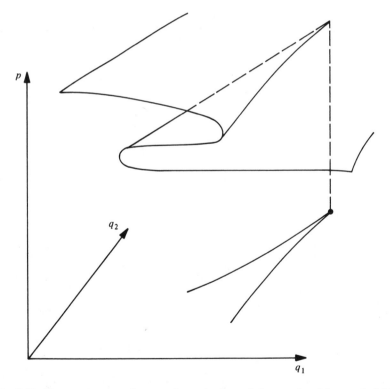

Fig. 6.12. Cusp points are the generic projection of the meeting of smooth folds in the phase space.

always one less than that of the configuration space; for example, folds will form two-dimensional surfaces in the case of a three-freedom system. For this reason folds are said to have *codimension* 1.

Caustic lines can join at a *cusp point* (fig. 6.12). Cusps are the only other generic caustic in the plane. Since their dimension is two less than the configuration space, they are of codimension 2. Hence, they appear along lines in a three-dimensional space. So far we have not met any cusp point, but these can easily be obtained by rotating the torus of fig. 6.6c by a group of symplectic transformations. The result is the same as the projection of a rotated tyre tube depicted in fig. 6.13. The numbers in the figure specify the multiplicity of $S(\mathbf{q})$ in each region.

Thom's theorem also furnishes the *normal form* to which we can reduce the action function $S(\mathbf{p})$ by a canonical transformation near the generic caustic. The most elegant presentation in the context of Hamiltonian dynamics is that of Arnold (1978, app. 12), who has considerably extended the classification of caustics. The idea is to use the mixed action function $S(p_1, \ldots, p_l, q_{l+1}, \ldots, q_L)$, which can be considered to be the Legendre transform of $S(\mathbf{q})$ with respect to l variables, or the transform of $S(\mathbf{q})$ with respect to $L - l$ variables. The transformation $\mathbf{p} \to \mathbf{q}$ is given implicitly by

$$q_i = \partial S/\partial p_i, \qquad p_j = \partial S/\partial q_j \qquad (6.39)$$

with $1 \leq i \leq l$ and $l < j \leq L$.

A nonsingular region has the normal form

$$A_1 : S = \sum_i p_i^2; \qquad (6.40)$$

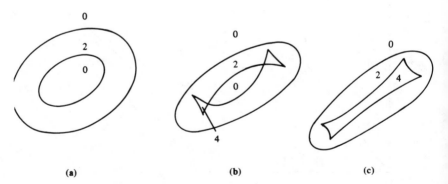

Fig. 6.13. Cusps appear in the shadows of a semitransparent tyre tube as it is rotated from a horizontal to a vertical position. The numbers in the figure indicate the multiplicity of the projection in each region.

that is, the Lagrangian manifold can be linearized in the absence of a caustic. For $L = 1$, the only generic singularity is the fold with the normal form

$$A_2 : S = \pm p_1^3. \tag{6.41}$$

For $L > 1$ we obtain the normal form for the fold line ($L = 2$), the fold surface ($L = 3$) and so on, simply by adding $L - 1$ terms of the normal form A_1 to A_2. This procedure for constructing the normal form for singularities in higher-dimensional spaces by adding A_1 is also valid for the singularities that follow. For $L = 2$ the other generic caustic is the cusp, the normal form of which is

$$A_3 : S = \pm p_1^4 + q_2 p_1^2. \tag{6.42}$$

For $L = 3$ we have two new generic codimension-3 caustics:

$$A_4 : S = \pm p_1^5 + q_3 p_1^3 + q_2 p_1^2, \tag{6.43}$$

$$D_4^\pm : S = \pm p_1^2 p_2 \pm p_2^3 + q_3 p_2^2. \tag{6.44}$$

For $L = 4$ there is also

$$A_5 : S = \pm p_1^6 + q_4 p_1^4 + q_3 p_1^3 + q_2 p_1^2, \tag{6.45}$$

$$D_5 : S = \pm p_1^2 q_2 \pm p_2^4 + q_4 p_2^3 + q_3 p_2^2, \tag{6.46}$$

and finally, for $L = 5$,

$$A_6 : S = \pm p_1^7 + q_5 p_1^5 + \cdots + q_2 p_1^2, \tag{6.47}$$

$$D_6 : S = \pm p_1^2 p_2 \pm p_2^5 + q_5 p_2^4 + q_4 p_2^3 + q_3 p_2^2, \tag{6.48}$$

$$E_6 : S = \pm p_1^3 \pm p_2^4 + q_5 p_1 q_2^2 + q_4 p_1 p_2 + q_3 p_2^2. \tag{6.49}$$

Arnold (1975) also studied the generic caustics for $L > 5$, but they cannot all be reduced to a discrete set.

The catastrophes $D_4^\pm$ (6.44) are known as the umbilics. If both the first terms have the same sign, we have the *hyperbolic umbilic* D_4^+; otherwise, the normal form represents the *elliptic* umbilic.

Exercises

1. Show that the transformation

$$p_1 = 2^{-2/3} 3^{1/2} (\mathfrak{p}_1 - \mathfrak{p}_2), \qquad p_2 = 2^{-2/3} (\mathfrak{p}_1 + \mathfrak{p}_2)$$

reduces the normal form for D_4^+ to

$$\pm \mathfrak{p}_1^3 \pm \mathfrak{p}_2^3 + 2^{-4/3} q_3 (\mathfrak{p}_1 + \mathfrak{p}_2)^2.$$

Find the transformation $\mathbf{q} \to \mathbf{q}$ that makes $(\mathbf{p}, \mathbf{q}) \to (\mathbf{p}, \mathbf{q})$ a canonical transformation.

2. Show that the folds of a separable system meet each other along the $q_3 = 0$ (i.e., $q_3 = 0$) section of the hyperbolic umbilic.

The double fold lines of a separable system do not meet at a cusp point, the only other generic singularity for a two-dimensional system. According to exercise 2 above, this meeting takes place in a section of a higher singularity. Thom's theorem thus leads us to expect that a small deformation of the torus will 'almost always' *unfold* this singularity into generic caustics. These are obtained by taking sections of D_4^+ with q_3 small but finite; that is, we consider q_3 not to be a coordinate in a higher-dimensional phase space but a parameter measuring the 'departure from separability'. The result of the unfolding is shown in fig. 6.14. The deformation of a separable torus in two dimensions may therefore have caustics resembling that of the torus with circular symmetry in fig. 6.13c. This is not the only possibility, however, since catastrophe theory provides only local results. Ozorio de Almeida and Hannay (1982) present transformations that bring separable tori into each of the configurations in fig. 6.15.

How sure can we be that a perturbation of the torus is generic – that is, capable of unfolding the projection of the torus? There will obviously be no unfolding if the perturbed Hamiltonian is still separable. What about an invariant torus of a general $\mathbf{p}^2/2m + V(\mathbf{q})$ type of Hamiltonian? Even in this case, it was shown by Ozorio de Almeida and Hannay (1982) that there is no unfolding of the D_4^+ corners because of the symmetry in the momenta. The unfoldings in fig. 6.15 are all realized by breaking this symmetry. The central problem in utilizing Thom's theorem or other arguments based on

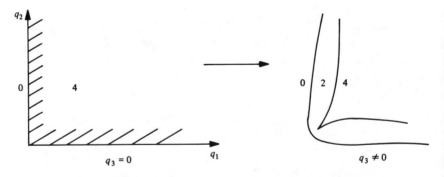

Fig. 6.14. The double fold lines of the hyperbolic umbilic are unfolded so that there are regions with zero, two or four pre-images on the torus.

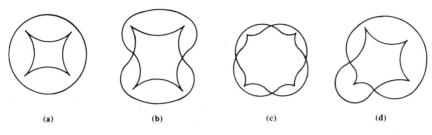

(a)   (b)   (c)   (d)

Fig. 6.15. Possible global unfoldings for a separable torus.

genericity or structural stability is to know whether the perturbation is really generic or special, owing to a possibly hidden symmetry.

The original concept of caustic is that of an envelope of a family of trajectories. In our case the family is the projection onto configuration space of the orbits that wind around an invariant torus. When the orbits pass smoothly from one torus branch to another, their projections touch a caustic. The simplest example of a caustic is that of a hose sprinkling parallel jets of water (fig. 6.16). The caustic is the horizontal line of maximum height attained by the water. It corresponds to a *singularity in the density of orbits*. If the orbits are identified with the light rays of optics, the resulting high intensity of light leads to 'burning' – hence the name. Berry and Upstill (1980) provide a clear presentation of caustics and catastrophe theory in an intuitive optical context.

In an integrable system where the frequencies do not have a constant ratio, almost all tori have open orbits that are dense. It follows that we can take a single orbit as 'the family' and obtain the caustic as the envelope of successive windings of this orbit. Furthermore, the integrability of the system can be ascertained if almost all orbits have caustic envelopes. As an example we first consider the circular billiard in fig. 6.17a. We know this to be an integrable system, because the force is central. Any orbit has a point of closest approach to the centre, at which it is tangent to an inscribed circle – the caustic. It is much harder to show directly that the elliptic billiard is an

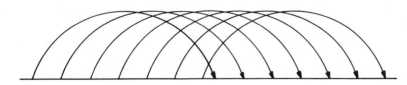

Fig. 6.16. One-dimensional water sprinkler. The horizontal line of highest ascent is an envelope of orbits, i.e., a caustic.

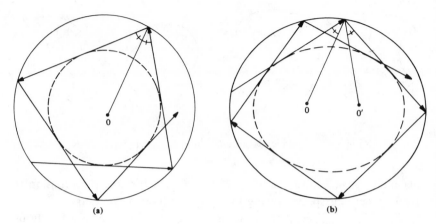

Fig. 6.17. (a) An orbit in a circular billiard has an inscribed circular envelope. The equality of the angles of incidence and reflection with respect to the radius leads to the conservation of angular momentum. (b) An orbit in an elliptic billiard also has an envelope – a confocal ellipse. The equality of the angles of incidence and reflection with respect to the lines to the foci at 0 and $0'$ leads to the conservation of the product of angular momenta about 0 and $0'$.

integrable system. However, in this case it is also true that any orbit has as envelope an inscribed curve, in this case a confocal ellipse (fig. 6.17b).

Exercise

Show that the product $J_0 J_{0'}$ of the angular momenta about the foci 0 and $0'$ is conserved for the circular billiard. (*Hint:* The angles between the tangent to an ellipse, at a given point, and the two lines joining it to the foci are equal.)[1]

6.4 Perturbations and averaging

The introduction to the theory of integrable systems presented in the preceding sections will serve as the basis for the traditional treatment of the semiclassical limit of quantum mechanics in chapter 7. The very particular restrictions on integrable systems prevent them from being structurally stable. Thus, in spite of the great number of solved integrable problems in classical mechanics, one should not expect that a general perturbation will lead to a new integrable system. Yet it is still very worthwhile to base the understanding of nearly integrable systems on the unperturbed system. In

---

[1] This result was communicated to me by J. H. Hannay.

section 6.5 we shall discuss the fundamental question of the survival of invariant tori subject to perturbations, the subject of the KAM theorem. We start with the discussion of approximate methods – that is, low-order perturbation theory and the averaging principle – which provide good quantitative results for limited time intervals.

It is convenient to study the flow of a perturbed system with the action-angle variables of the integrable system. The equations of motion for a quasi-integrable system have the form

$$\dot{\boldsymbol{\phi}} = \boldsymbol{\omega}_0(\mathbf{I}) + \varepsilon \mathbf{f}(\mathbf{I}, \boldsymbol{\phi}),$$

$$\dot{\mathbf{I}} = \varepsilon \mathbf{g}(\mathbf{I}, \boldsymbol{\phi}). \tag{6.50}$$

If the Hamiltonian has the form

$$H(\mathbf{I}) = H_0(\mathbf{I}) + \varepsilon H_1(\mathbf{I}, \boldsymbol{\phi}), \tag{6.51}$$

then

$$\mathbf{f} = \partial H_1/\partial \mathbf{I} \quad \text{and} \quad \mathbf{g} = -\partial H_1/\partial \boldsymbol{\phi}. \tag{6.52}$$

Consider 'disconnecting the perturbation' so that all the orbits are restricted to the invariant tori of H_0. We can then evaluate the time average of $\mathbf{g}(\mathbf{I}, \boldsymbol{\phi}(t))$. Almost all tori will have rationally independent frequencies, so the equivalence of time and configuration averages is valid:

$$\bar{\mathbf{g}}(\mathbf{I}) \equiv \frac{1}{T} \lim_{T \to \infty} \int_0^T \mathbf{g}(\mathbf{I}, \boldsymbol{\phi}(t)) \, dt = (2\pi)^{-L} \int_0^{2\pi} \cdots \int_0^{2\pi} \mathbf{g}(\mathbf{I}, \boldsymbol{\phi}) \, d\boldsymbol{\phi}. \tag{6.53}$$

In other words, the residue of the periodic force,

$$\tilde{\mathbf{g}}(\mathbf{I}, \boldsymbol{\phi}) \equiv \mathbf{g}(\mathbf{I}, \boldsymbol{\phi}) - \bar{\mathbf{g}}(\mathbf{I}), \tag{6.54}$$

has zero time average for periods T much larger than the characteristic periods $2\pi\omega_0^{-1}$ of the integrable system. This suggests that, if we now reconnect the perturbation so that the system moves according to (6.50), the effect of the force $\tilde{\mathbf{g}}$ will be small, provided that the 'inertia' ε^{-1} of the system is large and the unperturbed periods are small. The system is already receiving a contrary force before it has had time to respond to the original force. This argument leads to the approximation of (6.50) by the *averaged system*

$$\dot{\mathbf{I}}' = \varepsilon \bar{\mathbf{g}}(\mathbf{I}'). \tag{6.55}$$

This conjecture is confirmed for the following simple example presented by Arnold (1978, sec. 52):

$$\dot{\phi} = \omega \neq 0, \quad \dot{I} = \varepsilon g(\phi).$$

The solution is

$$I(t) - I(0) = \int_0^t \varepsilon g(\phi_0 + \omega t)\, dt = \int_0^t \varepsilon \bar{g}\, dt + \frac{\varepsilon}{\omega} \int_{\phi_0}^{\phi_0 + t/\omega} \tilde{g}(\phi)\, d\phi$$

$$= \varepsilon \bar{g} t + \frac{\varepsilon}{\omega} h\left(\frac{t}{\omega}\right),$$

where $h(\phi)$ is a periodic function and therefore bounded. Thus, if $I'(t)$ is the motion of the averaged system, we have

$$|I(t) - I'(t)| < C\varepsilon,$$

where $C = \|h\|\omega^{-1}$. The validity of the averaging principle for L-freedom systems, subject to a perturbation depending on a single angle under quite general conditions, is also proved by Arnold (1978, sec. 52); that is,

$$|I(t) - I'(t)| < C\varepsilon \qquad \text{for} \qquad 0 \le t \le \varepsilon^{-1}. \tag{6.56}$$

The averaging principle must be applied with care if the system has more than one freedom and the perturbation also depends on more than one angle. The problem resides in the tori with rationally dependent frequencies: An unperturbed periodic orbit will not average out a perturbation that depends on more than a single angle. If the frequency ratio ω_1/ω_2 is a low-order rational, the unperturbed periodic orbit will sample only an atypical part of $g(\phi_1, \phi_2)$, and even the open orbits of neighboring tori will take a long time to sweep through all the angles. We cannot apply the full averaging principle in such a region. Still it is possible to make a partial average along the direction of the unperturbed periodic orbit, the *fast variable*, while keeping the *slow variables* transverse to the motion generated by H_0.

A superior alternative to averaging away oscillatory terms of a Hamiltonian perturbation is to eliminate them by means of canonical transformations. We have already met a particular form of this procedure in chapter 4. There the unperturbed integrable Hamiltonian was quadratic, so that in the neighbourhood of the origin we could treat the higher-order angle-dependent terms in the Taylor expansion as the perturbation. We will now study the general perturbed Hamiltonian (6.51), following the presentation of Lichtenberg and Lieberman (1983).

The function H_1 is multiply periodic in the angles ϕ, so it can be developed in the Fourier series

$$H_1(\mathbf{I}, \phi) = \sum_{\mathbf{m}} H_{1\mathbf{m}}(\mathbf{I}) \exp(i\mathbf{m} \cdot \phi). \tag{6.57}$$

The Hamiltonian is then subject to a transformation $(\mathbf{I}, \boldsymbol{\phi}) \to (\mathbf{I}', \boldsymbol{\phi}')$, generated implicitly by

$$S(\mathbf{I}', \boldsymbol{\phi}) = \mathbf{I}' \cdot \boldsymbol{\phi} + \varepsilon S_1(\mathbf{I}', \boldsymbol{\phi}), \tag{6.58}$$

where

$$S_1(\mathbf{I}', \boldsymbol{\phi}) = \sum_{\mathbf{m}} S_{1\mathbf{m}}(\mathbf{I}') \exp(i\mathbf{m} \cdot \boldsymbol{\phi}). \tag{6.59}$$

Differentiating,

$$\mathbf{I} = \mathbf{I}' + \varepsilon \, \partial S_1(\mathbf{I}', \boldsymbol{\phi})/\partial \boldsymbol{\phi},$$
$$\boldsymbol{\phi}' = \boldsymbol{\phi} + \varepsilon \, \partial S_1(\mathbf{I}', \boldsymbol{\phi})/\partial \mathbf{I}', \tag{6.60}$$

we obtain the explicit transformation in the form

$$\mathbf{I} = \mathbf{I}' + \varepsilon \partial S_1(\mathbf{I}', \boldsymbol{\phi}')/\partial \boldsymbol{\phi} + O(\varepsilon^2),$$
$$\boldsymbol{\phi} = \boldsymbol{\phi}' - \varepsilon \partial S_1(\mathbf{I}', \boldsymbol{\phi}')/\partial \mathbf{I}' + O(\varepsilon^2). \tag{6.61}$$

The Hamiltonian is invariant with respect to this transformation, so

$$H'(\mathbf{I}', \boldsymbol{\phi}') = H_0(\mathbf{I}') + \varepsilon H_1(\mathbf{I}', \boldsymbol{\phi}') + \varepsilon \frac{\partial H_0}{\partial \mathbf{I}'} \cdot \frac{\partial S_1}{\partial \boldsymbol{\phi}'} + O(\varepsilon^2). \tag{6.62}$$

We can now specify the function S_1 so as to cancel the angular dependence of the linear part of H'. In this way the new Hamiltonian becomes integrable up to $O(\varepsilon^2)$. The required condition is

$$\boldsymbol{\omega} \cdot \frac{\partial S_1}{\partial \boldsymbol{\phi}'} = - \sum_{\mathbf{m} \neq 0} H_{1\mathbf{m}}(\mathbf{I}') \exp(i\mathbf{m} \cdot \boldsymbol{\phi}'), \tag{6.63}$$

which yields

$$S_1(\mathbf{I}', \boldsymbol{\phi}) = i \sum_{\mathbf{m} \neq 0} \frac{H_{1\mathbf{m}}(\mathbf{I}')}{\mathbf{m} \cdot \boldsymbol{\omega}} \exp(i\mathbf{m} \cdot \boldsymbol{\phi}). \tag{6.64}$$

Once again we have to face the problem of small denominators. Generally the frequencies vary continuously among the tori, so the *resonant tori*, for which

$$\mathbf{m} \cdot \boldsymbol{\omega}(\mathbf{I}) = 0, \tag{6.65}$$

are dense in phase space. We can study the first-order effect of the perturbation in the neighbourhood of a given resonant torus by adopting its periodic orbits as coordinates. In the case of two freedoms, the resonant torus will have frequencies

$$\omega_1/\omega_2 = s/r. \tag{6.66}$$

If we choose one of the angular coordinates to be

$$\phi = \phi_1 - s\phi_2/r, \tag{6.67}$$

then $\dot{\phi} = 0$ on the unperturbed torus. We need a full canonical transformation $(I_1, I_2, \phi_1, \phi_2) \rightarrow (J, I, \theta, \phi)$ incorporating (6.67). This is implicitly defined by the generating function

$$F = (\phi_1 - s\phi_2/r)I + (\phi_2/r)J, \tag{6.68}$$

so that the rest of the transformation becomes

$$\theta = \phi_2/r, \qquad I = I_1, \qquad J = rI_2 + sI_1. \tag{6.69}$$

Exercise

Find the irreducible circuits $\theta = 0$ and $\phi = 0$ on the torus and show that the respective symplectic areas are $2\pi J$ and $2\pi I$.

This exercise identifies the variables (J, θ) with the periodic orbit variables introduced in section 2.5. In the present integrable case, each orbit belongs to a family with two parameters – the energy and the angle ϕ.

$$H = H_0(J, I) + \varepsilon H_1(J, I, \theta, \phi), \tag{6.70}$$

where

$$H_1 = \sum_{k,m} H_{k,m}(J, I) \exp\{i[k\phi + (ks + mr)\theta]\}. \tag{6.71}$$

Though we cannot average over both angles, this is a reasonable approximation for the fast variable θ. The result is

$$\bar{H} = H_0(J, I) + \varepsilon \bar{H}_1(J, I, \phi), \tag{6.72}$$

where

$$\bar{H}_1 = \sum_{n=0}^{\infty} H_{-nr,ns}(J, I) \exp(-inr\phi). \tag{6.73}$$

By abandoning the attempt to eliminate all the angular dependence of the Hamiltonian, we thus obtain an improved approximation. The Hamiltonian $\bar{H}$ is integrable, since J is an independent constant of the motion.

Exercise

Derive the approximation (6.72) of the Hamiltonian (6.70) through perturbation theory and show that the generating function for the corresponding canonical transformation has no small denominators. Note that the identity of the two approximate Hamiltonians does not imply that the motion is the same.

If $\bar{H}_1$ is an analytic function (in a complex strip $|\mathrm{Im}\,\phi| < \rho$), then the coefficients in the Fourier series (6.73) decay exponentially. The simplest case is the one in which the only nonzero terms have $|n| \leq 1$. For small ε we can then make an approximate Taylor expansion about the unperturbed periodic torus with actions (I_0, J_0), to obtain

$$\bar{H} = H_0(I_0, J_0) + \varepsilon H_{0,0}(I_0, J_0) + a(\Delta I)^2 + \varepsilon b \cos(r\phi). \qquad (6.74)$$

This approximation is known as a *Chirikov resonance* (see Chirikov, 1979). We immediately recognize the pendulum Hamiltonian in the variables $(\Delta I, \phi)$ studied in section 4.4. There are r families of islands, corresponding to a single family of island tori. These are centred by a single stable periodic orbit, and they are separated from the enveloping tori by the separatrices that emanate from the unique unstable periodic orbit. The inclusion of further terms in the series (6.73) merely distorts the pendulum tori if all the higher-order terms are small. In contrast, the inclusion of θ dependent terms will generally lead to homoclinic intersections of the separatrices and hence to the existence of chaotic orbits. The appearance of pairs of stable and unstable periodic orbits at a resonance of a perturbed integrable system can be proved by topological arguments. This is the content of the Poincaré–Birkhoff theorem, the proof of which is sketched by Berry (1978).

Of course, we can apply the above theory again to the resonant island tori as in Lichtenberg and Lieberman (1983). These tori will break up into new islands and so on in an infinite hierarchy. The complexity of this picture leads to doubts that there is any truth in it. Do any tori survive the perturbation? This is the subject of the following section.

6.5 Discussion of the KAM theorem

A first approximation to the perturbed Hamiltonian (6.51) in the neighbourhood of a perturbed torus is given by (6.72) and (6.73). The appearance of unstable fixed points leads to motion that is radically different from that generated by the original Hamiltonian. Further perturbation from this resonant approximation still alters important features of the motion, but it generally cannot bring back even distortions of the unperturbed tori usually destroyed in the resonant region. This is because the unstable periodic orbits will show up as hyperbolic fixed points in a Poincaré section. These points with their separatrices are incompatible with the unperturbed torus family. Nonetheless, unstable fixed points are structurally stable within a one-parameter family of maps, as was shown in section 2.5.

Taking into account all the resonances up to a given *order* [defined as the

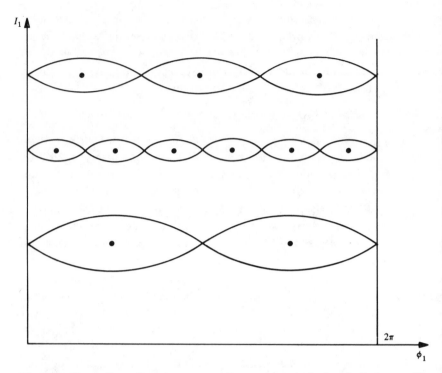

Fig. 6.18. Generally each region surrounding an unperturbed periodic torus will develop a resonance, but the resonant region becomes very narrow for those tori with very long periodic orbits.

biggest integer $|r|$ or $|s|$ in (6.66)], we obtain a picture such as fig. 6.18. The exponential decay of the coefficients $H_{k,m}$ in (6.71) implies that the size of the resonance – that is the area between the separatrices – proportional to εb in (6.74) decreases exponentially with the order of the resonance. We can thus obtain a first approximation to the flow as a whole by extending fig. 6.18 to all orders of resonance, while omitting those higher-order resonances that overlap and hence are 'swallowed up' by lower-order resonances. The resonant tori are densely distributed in the unperturbed system. Can tori survive the perturbation in the midst of all the resonances?

In the case of two freedoms, the Poincaré map for an unperturbed torus can be reduced to a translation on a circle. In chapter 5 we found that perturbations of this simple motion can still be reduced to it, provided that the rotation number is 'sufficiently irrational'. In the present context the rotation number can be identified with the frequency ratio, which varies continuously among the tori. Resonant tori correspond to circle maps whose rotation numbers are rational. Therefore, the basic theorem in section 5.4 proves directly that, if a torus 'sufficiently far' from resonance survives a perturbation, its Poincaré map can be reduced to a translation; that is, its flow will be quasi-periodic, just as in the integrable system.

The demonstration that tori do actually survive is much more intricate than the theory of the circle map. I will sketch only briefly the proof of Arnold (1963). The first step is to show the convergence of the Fourier series for the generating function (6.58) of the canonical transformation (I, ϕ) → (I', ϕ'), corresponding to first-order perturbation theory. To this end we assume that $H_1(I, \phi)$ is analytic, with $\| H_1 \| < \varepsilon$ in a strip $|\text{Im } \phi| < \rho$ (as with the circle map) and in a domain G of I. Convergence is proved in a reduced strip $|\text{Im } \phi'| < \rho - \delta$ and in a subdomain of G. This includes all the tori with actions I' and frequencies

$$\omega' = \partial H' / \partial I' \tag{6.75}$$

satisfying the Diophantine condition

$$|\omega' \cdot k| \geq c|k|^{-\nu}. \tag{6.76}$$

The constants c and ν depend only on ε and not on the integer vector k.

The result of this transformation is that the original Hamiltonian is reduced to the form

$$H'(I', \phi') = H'_0(I') + \varepsilon^2 H'_2(I', \phi') \tag{6.77}$$

along the tori satisfying (6.76). By applying Kolmogorov's method, as presented in section 5.4, it can now be shown that successive canonical transformations cancel the ϕ-dependent term while preserving a residual analyticity strip for the Hamiltonian. In this way the KAM theorem proves that the majority of the volume of phase space is occupied by invariant tori as $\varepsilon \to 0$. Moser's demonstration of the survival of invariant curves in area-preserving maps relaxes the condition of analyticity on H (Moser, 1962). On the other hand, Arnold's proof applies to perturbations of integrable systems with any number of freedoms.

The tori in the KAM theorem are specified by their frequencies (6.75) and (6.76). Therefore, the validity of the theorem depends on the nondegeneracy of the map $I \to \omega$; that is, we must add the condition that

$$\det |\partial^2 H / \partial I^2| \neq 0. \tag{6.78}$$

The KAM theorem can be adapted to integrable systems that are perturbed by a time-periodic Hamiltonian, since this can always be transformed into an autonomous system in an expanded phase space (see section 2.5). An important corollary for a forced system with one freedom is that linearly stable equilibria are truly stable, even after the perturbation. The reason is that near the equilibrium the Hamiltonian can be approximated by the truncated Birkhoff normal form, as in section 4.4. The remainder can be considered to be a perturbation that vanishes as the equilibrium is approached. The KAM theorem thus guarantees the existence of invariant curves surrounding the equilibrium. The orbits within are continuous and cannot intersect the curves. Since invariant tori exist arbitrarily close to the equilibrium, the motion in its neighbourhood is stable.

The motion for an autonomous system with two freedoms is restricted to three-dimensional energy surfaces. The KAM theorem guarantees the survival of small two-dimensional Birkhoff tori with the same energy as that of a point of linearly stable equilibrium. The equilibrium is inside any one of these tori on the energy shell. Thus, other orbits will remain within, even if they are not actually on a torus. The motion near a linearly stable periodic orbit of an autonomous two-freedom system is also stable, by a similar argument.

Exercise

Explain why we must invoke the Birkhoff normal form, rather than apply the KAM theorem directly to the integrable linearized system, by considering the whole nonlinear part as a perturbation.

The stability will be lost only when the last enveloping torus is destroyed. Presumably this torus is characterized by the frequency ratio most difficult to approximate by rationals. This conjecture has been computationally verified by Greene (1979) for the 'standard map'. This large perturbation regime is not accessible to the KAM techniques – hence, the great interest in the adaptation by Escande (1985) of 'renormalization group' techniques to the study of the survival of tori subject to large perturbations.

A relatively recent discovery of great importance by Aubry (1978) and Percival (1979) is that of nonchaotic, nonperiodic orbits resulting from the destruction of nonresonant tori by large perturbations. General existence proofs are given by Katok (1982), Mather (1982) and Aubrey and Le Daeron (1983). This orbit maintains the same rotation number as on the torus before destruction, but it does not fill a torus densely. There are gaps, creating a structure described as a product of a torus with a Cantor set. These so-called *cantori* hinder the diffusion of chaotic orbits, as studied by Mackay, Meiss and Percival (1984). Immediately beyond the destruction of a torus, holes of the cantorus are exponentially small, entailing exponentially slow diffusion through the gaps.

The situation for systems with three freedoms resembles that of the motion near cantori, even for arbitrarily small perturbations. The reason is that a three-dimensional torus does not separate the five-dimensional energy surface. It can be circumvented, just as a line in three-dimensional space. The Poincaré map for an integrable system will consist of linear translations along two-dimensional tori by angles $(2\pi\mu_1, 2\pi\mu_2)$. All the tori with rotation numbers satisfying

$$m_1\mu_1 + m_2\mu_2 = k \qquad (6.79)$$

are resonant. The KAM theorem guarantees the survival of only sufficiently nonresonant tori. These are the ones that are not caught in *Arnold's net*, partially sketched in fig. 6.19. Each band in the net corresponds to one of

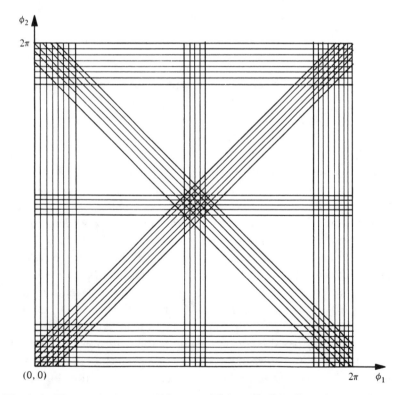

Fig. 6.19. The resonance condition establishes straight lines in the plane of rotation numbers (μ_1, μ_2) for the unperturbed system. The resonant region for each of these is thus a band. These are multiply connected in Arnold's net.

the resonance conditions (6.79). They become exponentially narrow as the order of the resonance becomes large. Instead of an infinite number of separated chaotic regions, as with two freedoms, we have a single chaotic region that penetrates everywhere. Since the bands communicate, an orbit started inside the chaotic region of one particular band will be able to wander around and explore all the other bands. The net is dense on the square, so the orbit will be able to go everywhere. This is known as *Arnold diffusion*. The diffusion will be exponentially slow as $\varepsilon \to 0$, because of the exponential narrowness of the higher-order bands. This subject is extensively reviewed by Lichtenberg and Lieberman (1983).

The proof in section 6.1 that the invariant tori of an integrable system are Lagrangian cannot be generalized to the survivors of a KAM perturbation. Nonetheless, this important property also holds for any invariant surface contained in an energy shell. This is a consequence of the same argument that led to the Poincaré–Cartan theorem in section 2.4. The full action

$$\oint [\mathbf{p} \cdot d\mathbf{q} - H dt] = 0 \qquad (6.80)$$

for any reducible closed circuit on the surface formed by a one-parameter family of trajectories in extended phase space. If all the orbits have the same energy, the loop can be projected into phase space, yielding

$$\oint \mathbf{p} \cdot d\mathbf{q} = 0. \qquad (6.81)$$

Thus, the invariant tori of a KAM system have caustics, just as do those of an integrable system. Moreover, their local features should have the generic properties allowed by catastrophe theory, unless there are special symmetries involved. Since all the invariant KAM tori have dense orbits, we can in principle determine whether an orbit belongs to a torus by simply graphing it in configuration space and watching out for envelopes. Figure 6.20 shows orbits computed by Ramaswamy and Marcus (1981) for a one-parameter family of quasi-integrable two-freedom Hamiltonians of the $\mathbf{p}^2 + V(\mathbf{q})$ type. The first six graphs indicate the existence of a torus. Note the nongeneric meeting of the folds at the umbilic corners, due to the symmetry in $\mathbf{p}$. The penultimate orbit is compatible with a thin island torus, since each thin ribbon shows a fold. The last graph indicates chaotic motion constrained by surviving tori: There are bounds to the motion, but clearly some orbits turn before reaching this 'outermost envelope'. The corresponding Poincaré sections are shown in fig. 6.21.

In a quasi-integrable system, we can be sure that an envelope for the successive windings of a trajectory is the signature of a torus. However, a more general system may have invariant surfaces with different topology. An example is provided by the 'pseudointegrable' systems studied by Richens and Berry (1981), which appear among polygon billiards. Among these, a two-dimensional invariant surface may be a g-handled sphere with $g > 1$ (g is called the *genus* of the surface). Ozorio de Almeida and Hannay (1982) present a rule for identifying g from the caustics of a two-dimensional invariant surface: The tangent to a smooth closed line swings through a multiple of 2π (it may self-intersect). If the closed line has cusps, however, it may swing through half-turns; that is, we consider the direction of the tangent line as that of a car travelling along a fold line, which backs away when it meets a cusp point. Then, if all the caustic lines are traversed the "same" way – that is, with the double layer of the fold always on the right (or always on the left) – *the total number of rotations of the tangent line is* $g - 1$. For a torus with umbilic corners, because of symmetry, the rule must be applied to the unfolded projection, as in fig. 6.15.

We have now completed the outline of classical dynamics. Many subjects omitted or merely touched upon are thoroughly discussed by Lichtenberg and Lieberman (1983), who also provide extensive references. My emphasis has been mainly on periodic orbits. In contrast to a dissipative system, where periodic orbits may be attractors, there is zero probability of the system being on a periodic orbit. However, we have seen many systems where periodic orbits are at least approximately dense. This is the case of integrable systems, simply because commensurable frequency ratios are dense among the real numbers. Perturbing these systems we generally destroy all the periodic tori. Yet isolated periodic orbits remain, half of which are stable. Around these we find nonresonant island tori, the periodic tori giving way to isolated periodic orbits and so on, in an ever finer self-similar structure. The complement of this figure is organized around the unstable periodic orbits, also present in the resonant region. Their separatrices generally have homoclinic or heteroclinic intersections – points of accumulation for infinite families of periodic orbits. Increasing the perturbation may bring about the destruction of most island tori and the period-doubling bifurcation cascades of the stable fixed points. The result is the population of most of phase space by a multitude of unstable periodic orbits, as in the cat map, where they are proved to be dense.

The hypothesis that periodic orbits are dense means that all orbits can be well approximated by periodic orbits for arbitrarily long times. One of the guiding principles for research in this field since Poincaré has thus been the parallel studies of the local motion near periodic orbits and of their

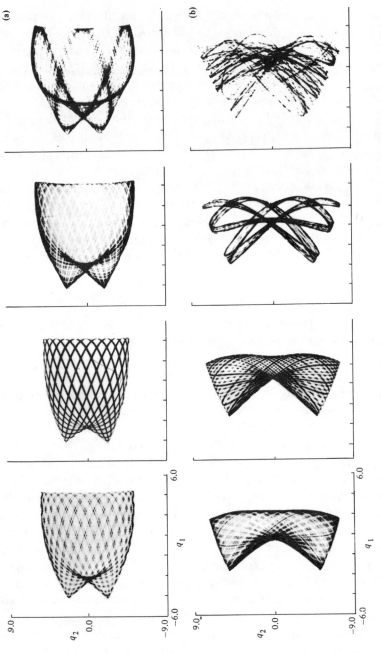

Fig. 6.20. Coordinate space orbits for a quasi-integrable system. (Courtesy of R. Ramaswamy and R. A. Marcus.)

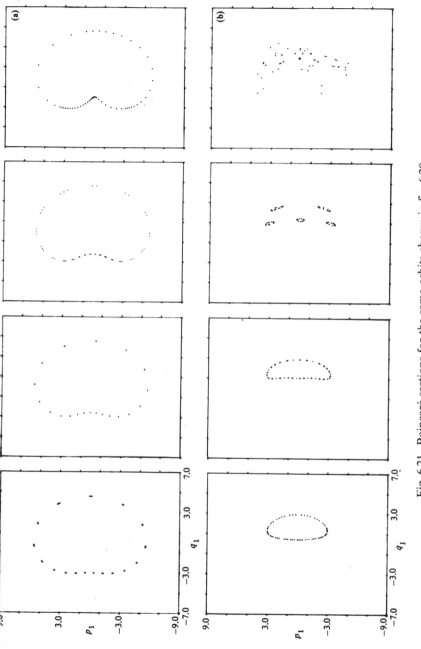

Fig. 6.21. Poincaré sections for the same orbits shown in fig. 6.20.

organization in phase space. The latter program is certainly at a much more incipient stage. All the same, it will be shown in chapter 9 that the same principle is also a valuable guide to the study of the semiclassical limit of quantum mechanics, our concern from here on.

Part II
Quantum dynamics

7

Torus quantization

In principle all the classical Hamiltonian systems studied in the first part of this book are only approximations to the systems of quantum mechanics. There are no unique rules for ascribing a quantum Hamiltonian to a known classical system. However, the inverse problem is perfectly well defined: Consider a quantum Hamiltonian $\hat{H} = H(\hat{p}, \hat{q})$, a function of the *position operators* $\hat{\mathbf{q}} \equiv (\hat{q}_1, \ldots, \hat{q}_L)$ and the *momentum operators* $\hat{\mathbf{p}} \equiv (\hat{p}_1, \ldots, \hat{p}_L) = -i\hbar \, \partial/\partial\mathbf{q}$, where $\hbar$ is *Planck's constant*; then the corresponding classical Hamiltonian is simply $H(\mathbf{p}, \mathbf{q})$. Much more is understood about the classical motion than about the corresponding quantum system. In the *semiclassical limit* $\hbar \to 0$ (i.e., where the ratio of $\hbar$ to other relevant parameters of the motion is small), we can apply our understanding of classical dynamics to the study of wave functions, energy levels and other quantum mechanical entities. The behaviour that emerges must be distinguished from the strict *classical limit*, $\hbar = 0$, because of the nonanalyticity of quantum mechanics at $\hbar = 0$.

The next three chapters are dedicated to the study of autonomous Hamiltonian systems. This chapter presents mainly the traditional semi-classical theory of the quantization of Lagrangian surfaces, which goes back to the work of Dirac, Van Vleck, Einstein, Brillouin, Keller and Maslov. The stationary states correspond to the invariant tori of classically integrable systems. This identification becomes even tighter in the Wigner–Weyl phase space representation, which is also discussed. The time evolution of an open Lagrangian surface corresponds to a nonstationary state, with which we start our study. A particularly important example is the semiclassical propagator, to be used as a tool for the study of nonintegrable systems in chapter 9.

7.1 Semiclassical wave functions

The *Schrödinger equation* determines the evolution of the wave function $\langle \mathbf{q} | \psi(t) \rangle$ generated by the Hamiltonian $H(\hat{\mathbf{p}}, \mathbf{q})$ in the coordinate representation,

$$ih\frac{\partial}{\partial t}\langle \mathbf{q}|\psi(t)\rangle = H(\hat{\mathbf{p}},\mathbf{q})\langle \mathbf{q}|\psi(t)\rangle, \qquad (7.1)$$

where $\hat{\mathbf{p}} = -i\hbar\,\partial/\partial\mathbf{q}$. The simplest case is that of a free particle, for which

$$H(\hat{\mathbf{p}},\mathbf{q}) = \hat{\mathbf{p}}^2/2m. \qquad (7.2)$$

The solutions are then *plane waves*

$$\langle \mathbf{q}|\psi_{\mathbf{p}}(t)\rangle = \exp\{i\hbar^{-1}(\mathbf{p}\cdot\mathbf{q} - Et)\}, \qquad (7.3)$$

where $E = \mathbf{p}^2/2m$ is the particle energy. The plane wave (7.3) is the *eigenfunction* of the momentum operator,

$$\hat{\mathbf{p}}\langle \mathbf{q}|\psi_{\mathbf{p}}(t)\rangle = \mathbf{p}\langle \mathbf{q}|\psi_{\mathbf{p}}(t)\rangle, \qquad (7.4)$$

and its wavelength $\lambda = 2\pi\hbar/|\mathbf{p}| \to 0$, as $\hbar \to 0$, for fixed $\mathbf{p}$.

Let us consider now a general classical Hamiltonian $H(\mathbf{p},\mathbf{q})$ in the neighbourhood of the coordinate $\mathbf{q} = \mathbf{q}_0$. The error in approximating the Hamiltonian by $H(\mathbf{p},\mathbf{q}_0)$ becomes arbitrarily small as this neighbourhood is reduced. However, it still contains many wavelengths of the corresponding plane wave solutions of the Schrödinger equation in the limit of $\hbar \to 0$. One way of obtaining the semiclassical limit of a general wave function is thus to subdivide the Hamiltonian in J steps, centred on coordinates $\mathbf{q}_j$, where $H = H(\mathbf{p},\mathbf{q}_j)$. The solutions of the corresponding Schrödinger equation will be plane waves having $\mathbf{p} = \mathbf{p}_j$, such that $H(\mathbf{p}_j,\mathbf{q}_j) = E$, a constant. The problem is reduced in this way to matching the amplitude and phase of the plane waves at the common boundaries. This procedure is feasible and useful for one-dimensional problems, as reviewed by Berry and Mount (1972).

In general it is simpler to note that in the limit $J \to \infty$ we obtain a wave function whose exponent is no longer linear in $\mathbf{q}$ and that the amplitude is now a function of $\mathbf{q}$. We therefore introduce the trial solution

$$\langle \mathbf{q}|\psi(t)\rangle = A(\mathbf{q},t)\exp[i\hbar^{-1}\sigma(\mathbf{q},t)] \qquad (7.5)$$

into the Schrödinger equation (7.1), as in Dirac (1930). The resulting equation is

$$i\hbar\frac{\partial A}{\partial t} - A\frac{\partial\sigma}{\partial t} = \exp(-i\hbar^{-1}\sigma)H(\hat{\mathbf{p}},\mathbf{q})\exp(i\hbar^{-1}\sigma)A. \qquad (7.6)$$

The right side divided by A is the coordinate representation of the operator

$$\exp[-i\hbar^{-1}\sigma(\hat{\mathbf{q}},t)]H(\hat{\mathbf{p}},\hat{\mathbf{q}})\exp[i\hbar^{-1}\sigma(\hat{\mathbf{q}},t)] = H(\hat{\mathbf{p}} + \partial\sigma/\partial\hat{\mathbf{q}},\hat{\mathbf{q}}). \qquad (7.7)$$

This is obtained by noting that $\exp[-i\hbar^{-1}\sigma(\hat{\mathbf{q}}, t)]$ is a unitary operator for which

$$\exp[-i\hbar^{-1}\sigma(\mathbf{q}, t)]\hat{\mathbf{p}}\exp[i\hbar^{-1}\sigma(\hat{\mathbf{q}}, t)] = \hat{\mathbf{p}} + \partial\sigma/\partial\mathbf{q} \qquad (7.8)$$

and that algebraic relations are preserved by unitary transformations. Thus, the Schrödinger equation takes on the exact form

$$i\hbar\frac{\partial A}{\partial t} - A\frac{\partial \sigma}{\partial t} = H\left(\hat{\mathbf{p}} + \frac{\partial \sigma}{\partial\mathbf{q}}, \mathbf{q}\right)A. \qquad (7.9)$$

If we now neglect all the terms containing $\hbar$, as a first approximation, there results

$$H\left(\frac{\partial \sigma}{\partial\mathbf{q}}, \mathbf{q}\right) + \frac{\partial \sigma}{\partial t} = 0. \qquad (7.10)$$

This is immediately recognised as the *Hamilton–Jacobi equation* of classical mechanics (Arnold, 1978; Goldstein, 1980). Given any initial condition $\sigma(\mathbf{q}, t_0) = S(\mathbf{q})$, we obtain the solution as the action

$$\sigma(\mathbf{q}, t) = S(\mathbf{q}_0) + \int_{\mathbf{q}_0, t_0}^{\mathbf{q}, t} L(\mathbf{q}, \dot{\mathbf{q}}, t)\, dt$$

$$= S(\mathbf{q}_0) + \int_{\mathbf{q}_0, t}^{\mathbf{q}, t} \{\mathbf{p}\cdot\dot{\mathbf{q}} - H\}\, dt, \qquad (7.11)$$

where $L(\mathbf{q}, \dot{\mathbf{q}}, t)$ is the *Lagrangian* of the system and the integration is carried out along the classical trajectories of $(\mathbf{q}, t)$ space. The evolution of the surface

$$\mathbf{p}(\mathbf{q}, t) = \partial\sigma/\partial\mathbf{q} \qquad (7.12)$$

in phase space corresponds to the semiclassical evolution of $\langle\mathbf{q}|\psi(t)\rangle$. This surface is Lagrangian at any time, that is,

$$\oint \mathbf{p}\cdot d\mathbf{q} = 0 \qquad (7.13)$$

for any reducible closed loop contained in it.

We seek stationary solutions analogous to the plane waves (7.3). Then $\mathbf{p}(\mathbf{q})$ will be independent of t and so

$$\partial\sigma/\partial t = -H(\mathbf{p}, \mathbf{q}) = -E, \qquad (7.14)$$

a constant, that is,

$$\sigma = S(\mathbf{q}) - E(t - t_0). \qquad (7.15)$$

The function $S(\mathbf{q})$ satisfies the equation

$$H(\partial S/\partial \mathbf{q}, \mathbf{q}) = E, \tag{7.16}$$

also called the (stationary) Hamilton–Jacobi equation. In the simple case of the free particle with the Hamiltonian (7.2) and one freedom, the solution is

$$S = \pm \int_{q_0}^{q} (2mE)^{1/2}\, dq = \pm (2mE)^{1/2}(q - q_0), \tag{7.17}$$

so the semiclassical solution is then exact.

We can consider the passage from (7.10) to (7.16) to be another instance of the reduction of the Hamiltonian by one freedom, due to the conservation of energy, as in section 2.5. This interpretation is made explicit by bringing forth the implicit function

$$\frac{\partial S}{\partial q_1} = - K\left(\frac{\partial S}{\partial \mathbf{Q}}, \mathbf{Q}; q_1\right), \tag{7.18}$$

where $\mathbf{Q}$ are the remaining $L-1$ coordinates, and interpreting K as the reduced Hamiltonian and $-q_1$ as the 'time'. This has the same form as (7.10), so any function $S_0(\mathbf{Q}_0)$ will give rise to a solution $S(\mathbf{Q}; q_1)$, generated by the trajectories of the reduced Hamiltonian according to (7.11). The problem is that, in contrast to time, the coordinate q_1 will generally wind back over itself, if $H(\mathbf{p}, \mathbf{q}) = E$ is a compact surface. If we attempt to extend the local solution obtained from an arbitrary $S_0(\mathbf{Q}_0)$, the 'returning solution' $S(\mathbf{Q}; q_1)$ will not usually match smoothly back onto $S_0(\mathbf{Q})$.

The search for global solutions to (7.16) in the end falls back on the problem of the existence of *invariant* Lagrangian surfaces for a given Hamiltonian. As we saw in chapter 6, these surfaces 'foliate' the entire phase space of an integrable system. They are guaranteed to exist in quasi-integrable systems, yet they are completely absent from ergodic systems and from the chaotic regions of quasi-integrable systems. In the absence of invariant surfaces, the stationary wave function cannot have the simple form (7.5). Its general features are still a substantially open problem.

We shall now determine the amplitude A of the wave function. Van Vleck (1928) derived it from the next term of an $\hbar^n$ expansion of equation (7.9). However, he also obtained it from a simple argument based on the uncertainty principle, which we here adopt. Let us consider the Lagrangian surface $\mathbf{p}(\mathbf{q})$ as being part of a family. The standard case is that of a family of tori (which need not be invariant) $\mathbf{p}(\mathbf{q}, \mathbf{I})$, so we shall use the corresponding notation. We can then consider that the function $\sigma(\mathbf{q}, \mathbf{I}, t)$ [or $S(\mathbf{q}, \mathbf{I})$]

generates a canonical transformation, defined by

$$\partial\sigma/\partial\mathbf{q} = \mathbf{p}, \qquad \partial\sigma/\partial\mathbf{I} = \boldsymbol{\phi}. \tag{7.19}$$

In the semiclassical limit, we can take $\mathbf{I}$ and $\boldsymbol{\phi}$ to correspond to quantum variables $\hat{\mathbf{I}}$ and $\hat{\boldsymbol{\phi}}$, with commutators

$$[\hat{I}_i, \hat{I}_j] = [\hat{\phi}_i, \hat{\phi}_j] = 0, \qquad [\hat{\phi}_i, \hat{I}_j] = i\hbar. \tag{7.20}$$

But $\mathbf{I}$ is a constant of the motion – it is known exactly. So in the $\boldsymbol{\phi}$ representation the wave function $\langle \boldsymbol{\phi} | \psi_\mathbf{I} \rangle$ must have a constant amplitude. From the conservation of probability under a change of variables,

$$|\langle \mathbf{q} | \psi_\mathbf{I} \rangle|^2 \, d\boldsymbol{\phi} = |\langle \mathbf{q} | \psi_\mathbf{I}(t) \rangle|^2 \, d\mathbf{q}, \tag{7.21}$$

we thus obtain

$$A^2 = |\psi_\mathbf{I}(\mathbf{q})|^2 \propto \left| \det \frac{\partial\boldsymbol{\phi}}{\partial\mathbf{q}} \right| = \left| \det \frac{\partial^2\sigma}{\partial\mathbf{q}\,\partial\mathbf{I}} \right|. \tag{7.22}$$

The semiclassical wave functions corresponding to a family of Lagrangian surfaces with actions $\sigma(\mathbf{q}, \mathbf{I}, t)$ therefore have the form

$$\langle \mathbf{q} | \psi_I(t) \rangle = c \left| \det \frac{\partial^2\sigma}{\partial\mathbf{q}\,\partial\mathbf{I}} \right|^{1/2} \exp\{i\hbar^{-1}\sigma(\mathbf{q}, \mathbf{I}, t)\}. \tag{7.23}$$

7.2 Wave functions for invariant tori

Tori are invariant Lagrangian surfaces, but we found in chapter 6 that they are specified by a multivalued action $S_j(\mathbf{q})$. None of the layers is invariant on its own, though each of the actions $S_j(\mathbf{q})$ satisfies the stationary Hamilton–Jacobi equation (7.16). Therefore, the corresponding solution of the time-independent Schrödinger equation is given by the superposition principle as

$$\langle \mathbf{q} | \psi_\mathbf{I} \rangle = c \sum_j \left| \det \frac{\partial^2 S_j}{\partial\mathbf{q}\,\partial\mathbf{I}} \right|^{1/2} \exp\{i\hbar^{-1}S_j(\mathbf{q}, \mathbf{I}) + i\alpha_j\}. \tag{7.24}$$

The common real factor c arises from the fact that the probability (7.21) is constant for each of the ranges of angles $\boldsymbol{\phi}$, corresponding to each of the layers of the torus. The problem with this semiclassical wave function is that it diverges at the boundaries of the torus layers. By resolving this difficulty, we shall also determine the constant phases α_j.

To start with, consider the simple one-freedom Hamiltonian of the form $H = p^2/2m + V(q)$. The torus (the invariant closed curve) is a symmetric function of p. The actions for both layers have the form

$$S_{\pm}(q, I) = \pm \int_{q_0}^{q} dq \{2m[E(I) - V(q)]\}^{1/2}. \tag{7.25}$$

Use of the chain rule and the fact that dE/dI, obtained from (6.20), is constant with the same value along both layers takes (7.24) into the form

$$\langle q|\psi_I \rangle = c \left| \frac{dE}{dI} \frac{1}{2} \{2m[E - V(q)]\}^{-1/2} \right|^{1/2} \{\exp(i\hbar^{-1}S_{+} + i\alpha_{+})$$
$$+ \exp(i\hbar^{-1}S_{-} + i\alpha_{-})\}. \tag{7.26}$$

At the turning point $E = V(q)$, the wave function diverges, though it has a finite integral. This wave function is identical with the Wentzel–Kramers–Brillouin approximation (see, e.g., Berry and Mount, 1972).

For a general torus wave function, the squared amplitude is given by

$$\det \frac{\partial^2 S}{\partial \mathbf{q} \, \partial \mathbf{I}} = \det \frac{\partial \boldsymbol{\phi}}{\partial \mathbf{q}} = \det \frac{\partial \boldsymbol{\phi}}{\partial \mathbf{p}} \det \frac{\partial \mathbf{p}}{\partial \mathbf{q}}. \tag{7.27}$$

At a caustic of the torus, the last determinant in (7.27) blows up, as discussed in section 6.3. This is a defect of the semiclassical wave function, but it is important to note again that the probability integral remains finite and the peak in the density reflects a real though finite increase in the probability amplitude near the caustic.

One way around this singularity of the wave function is to use the momentum representation. The semiclassical wave function is then

$$\langle \mathbf{p}|\psi_I \rangle = c \sum_{j} \left| \det \frac{\partial^2 S_j(\mathbf{p})}{\partial \mathbf{p} \, \partial \mathbf{I}} \right|^{1/2} \exp \{-i\hbar^{-1}S_j(\mathbf{p}, \mathbf{I}) + i\beta_j\}, \tag{7.28}$$

where the action

$$S_j(\mathbf{p}) = \int_{\mathbf{p}_0}^{\mathbf{p}} \mathbf{q}_j(\mathbf{p}) \cdot d\mathbf{p} \tag{7.29}$$

is the Legendre transform of $S_j(\mathbf{q})$, as in section 6.3.

Exercise

Derive (7.28) directly as the stationary semiclassical approximation to the solution of the equation

$$H(\mathbf{p}, i\hbar \, \partial/\partial \mathbf{p}) \langle \mathbf{p}|\psi \rangle = i\hbar \, \partial \langle \mathbf{p}|\psi \rangle / \partial t$$

using the trial solution

$$\langle \mathbf{p}|\psi(t) \rangle = A \exp \{i\hbar^{-1}[-S(\mathbf{p}) - Et]\}.$$

We found in section 6.3 that $\mathbf{q}(\mathbf{p})$ is also multivalued for an invariant torus and that the different layers join up exactly at the locus of $|\det \partial^2 S_j / \partial \mathbf{p}\, \partial \mathbf{I}| = \infty$. But these $\mathbf{p}$ caustics never coincide with the $\mathbf{q}$ caustics. In the regions free of any kind of caustic, both representations should be equivalent.

They are, of course, related by the Fourier transform

$$\langle \mathbf{q} | \psi \rangle = (2\pi\hbar)^{-L/2} \int d\mathbf{p} \langle \mathbf{p} | \psi \rangle \exp(i\hbar^{-1}\mathbf{p}\cdot\mathbf{q}). \qquad (7.30)$$

Introducing (7.28) into (7.30), we obtain a sum of integrals. Each integrand has a classical amplitude $|\det \partial^2 S_j / \partial \mathbf{p}\, \partial \mathbf{I}|^{1/2}$, independent of $\hbar$, multiplied by an oscillatory term. Any region of $\mathbf{p}$ with high-frequency oscillations will give a small contribution to the integral. This is dominated by the neighbourhoods of

$$(\partial/\partial\mathbf{p})[-S_j(\mathbf{p},\mathbf{I}) + i\hbar\beta_j + \mathbf{p}\cdot\mathbf{q}] = -\mathbf{q}_j(\mathbf{p}) + \mathbf{q} = 0. \qquad (7.31)$$

Since the frequencies of oscillation are proportional to $\hbar^{-1}$, we can approximate each integral by the contributions of small neighbourhoods of the *stationary points* in the limit $\hbar \to 0$. According to (7.31) these are exactly the points on the torus $\mathbf{p}_j(\mathbf{q})$ that project down onto the point $\mathbf{q}$. The simplest case occurs when all the stationary points are separated from one another by many wavelengths; precisely the case in which all caustics are far away. Then the action in the neighbourhood of each stationary point can be approximated by

$$S_j(\mathbf{p},\mathbf{I}) \simeq S_j(\mathbf{p}_j,\mathbf{I}) + \frac{1}{2}(\mathbf{p}-\mathbf{p}_j)\frac{\partial^2 S_j}{\partial\mathbf{p}^2}(\mathbf{p}-\mathbf{p}_j). \qquad (7.32)$$

The substitution of (7.32) for the exponent in the integrals (7.30) constitutes a good approximation in the dominant regions, while for the rest of the domain of integration the approximate integrals will also be highly oscillatory.

Exercises

1. Show that

$$\int_{-\infty}^{\infty} dp \exp\left(\frac{i\hbar^{-1}ap^2}{2}\right) = \left(\frac{2\pi\hbar}{a}\right)^{1/2} \exp\left(\frac{i\pi\gamma_a}{4}\right), \qquad (7.33)$$

where $\gamma_a = a/|a|$ is the sign of a.

2. A symmetric $L \times L$ matrix can be diagonalized by an orthogonal

transformation with unit Jacobian determinant. Hence, show that

$$\int_{-\infty}^{+\infty} d\mathbf{p}\exp\{i(2\hbar)^{-1}\mathbf{p}\mathscr{A}\mathbf{p}\} = \left(\frac{(2\pi\hbar)^L}{|\det\mathscr{A}|}\right)^{1/2}\exp\left(\frac{i\pi\gamma_{\mathscr{A}}}{4}\right), \qquad (7.34)$$

where $\gamma_{\mathscr{A}}$ is the signature of $\mathscr{A}$, that is, the number of positive minus the number of negative eigenvalues of this matrix.

The *stationary-phase approximation* consists of the substitution of the phase $\hbar^{-1}S_j$ by its quadratic approximation (7.31), while the relatively slowly varying amplitude is kept constant. Using (7.34), we obtain

$$(2\pi\hbar)^{-L/2}\int d\mathbf{p}\left|\det\frac{\partial^2 S_j}{\partial\mathbf{p}\,\partial\mathbf{I}}\right|^{1/2}\exp\{-i\hbar^{-1}[\mathbf{p}\cdot\mathbf{q}-S_j(\mathbf{p},\mathbf{I})]+i\beta_j\}$$

$$=\left|\det\frac{\partial^2 S_j}{\partial\mathbf{p}\,\partial\mathbf{I}}\right|^{1/2}\left|\det\frac{\partial^2 S_j}{\partial\mathbf{p}^2}\right|^{-1/2}\exp\{i\hbar^{-1}[\mathbf{p}_j(\mathbf{q})\cdot\mathbf{q}-S_j(\mathbf{p}_j(\mathbf{q}),\mathbf{I})]+i\alpha_j\}.$$

$$(7.35)$$

This expression is simplified by means of the identification

$$\mathbf{p}_j(\mathbf{q})\cdot\mathbf{q}-S_j(\mathbf{p}_j(\mathbf{q}),\mathbf{I})=S_j(\mathbf{q}), \qquad\qquad (7.36)$$

the Legendre transform. For the amplitude we have

$$\left|\det\frac{\partial\mathbf{q}}{\partial\mathbf{I}}\det{}^{-1}\frac{\partial\mathbf{q}}{\partial\mathbf{p}}\right|^{1/2}=\left|\det\frac{\partial\mathbf{q}}{\partial\mathbf{I}}\det\frac{\partial\mathbf{p}}{\partial\mathbf{q}}\right|^{1/2}=\left|\det\frac{\partial\mathbf{p}}{\partial\mathbf{I}}\right|^{1/2}=\left|\det\frac{\partial^2 S_j}{\partial\mathbf{q}\,\partial\mathbf{I}}(\mathbf{q},\mathbf{I})\right|^{1/2}.$$

$$(7.37)$$

The end result of taking the Fourier transform of $\langle\mathbf{p}|\psi_{\mathbf{I}}\rangle$ in a region without caustics within the stationary-phase approximation is thus the semiclassical wave function $\langle\mathbf{q}|\psi_{\mathbf{I}}\rangle$. This consistency of the coordinate and the momentum representation, where both are nonsingular, allows us to *define* the semiclassical wave function $\langle\mathbf{q}|\psi_{\mathbf{I}}\rangle$ near a caustic by the Fourier transform (7.30). This manner of avoiding semiclassical singularities is known as *Maslov's method*. Its equivalent in optics was discovered by Keller (1958). Maslov's treatment (see Maslov and Fedorink, 1981) is considerably more abstract than this presentation, which follows Berry (1981a).

Two or more points $\mathbf{p}(\mathbf{q})$ coalesce as one approaches a $\mathbf{q}$ caustic of the torus. The momenta of these points are defined by the stationary-phase condition (7.31). But the condition for the validity of the stationary-phase method is just that the stationary points be separated by many wavelengths, so this method cannot be applied close to a caustic. However, it is still true that the integral is dominated by the neighbourhoods of the points with stationary phase.

What we now need is an improvement of the too simple quadratic approximation to the action (7.32). This is readily supplied by the normal forms for the caustics presented in section 6.3. There exists a canonical transformation $(\mathbf{p}, \mathbf{q}) \rightarrow (\mathbf{p}, \mathbf{q})$, such that $S(\mathbf{p})$ is just one of the forms enumerated, depending on the local type of caustic. The wave function for $L = 1$ corresponding to the normal form of the action S_{A_j}, (6.41), (6.42), (6.43),..., can be modelled by

$$\langle q | \psi_{A_j} \rangle = \int_{-\infty}^{\infty} dp \exp\{i[qp - S_{A_j}(p)]\}, \qquad (7.38)$$

known as a *diffraction catastrophe integral*. Analogous definitions hold for the other normal forms D_j, and so on. For $L > 1$ we simply multiply $\langle q | \psi_{A_j} \rangle$ by

$$\langle \mathbf{Q} | \psi_{A_2} \rangle = \int_{-\infty}^{\infty} d\mathbf{P} \exp\{i(\mathbf{Q} \cdot \mathbf{P} \mp \mathbf{P}^2)\}; \qquad (7.39)$$

that is, the other degrees of freedom will have normal coordinates where the action is just a Gaussian. Reduction to (7.39) thus amounts to performing a stationary-phase integration in the $\mathbf{P}$ variables 'transverse to the caustic'.

The simplest case is that of the fold A_3, for which $S_{A_3}(p) = -p^3$. The stationary points $p^{\pm}(q)$ meet at $p = 0$ (fig. 7.1). The catastrophe integral has the standard form

$$\langle \mathbf{q} | \psi_{A_3} \rangle = \int_{-\infty}^{\infty} dp \exp\{i(p^3 + qp)\} = 2\pi 3^{-1/3} Ai(3^{-1/3}q), \qquad (7.40)$$

where $Ai(x)$ is the Airy function (Abramowitz and Stegun, 1964). Consider the family of curves for which the action is

$$S(p, I) = -p^3/3 + pI. \qquad (7.41)$$

The curves $q(p, I) = I - p^2$ are parabolas, so the parameter I (not an action variable here) distinguishes each parabola by the point where it crosses the q axis. The action measured from this point is

$$S_{\pm}(q, I) = \pm \int_{I}^{q} \sqrt{I - q}\, dq = \tfrac{2}{3}(I - q)^{3/2}, \qquad (7.42)$$

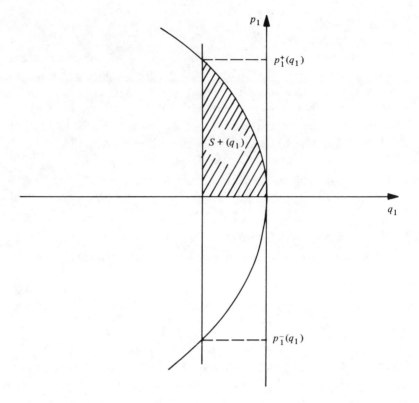

Fig. 7.1. Normal form for the fold line $S_{A_3}(p) = -p^3$. The stationary points $p^{\pm}(q)$ meet at $p = 0$.

and the corresponding wave function is

$$\langle q | \psi_I \rangle = c(2\pi\hbar)^{-1/2} \int_{-\infty}^{\infty} dp \exp\left\{ i\hbar^{-1}\left[\frac{p^3}{3} + (q - I)p \right] \right\}$$
$$= c(2\pi\hbar^{-1/3})^{1/2} Ai[\hbar^{-2/3}(q - I)]. \tag{7.43}$$

The maximum amplitude of $\langle q | \psi_I \rangle$ does not fall exactly on the caustic $(q = I)$, as can be seen from the graph of the Airy function in fig. 7.2. However, this maximal point tends to the caustic as $\hbar \to 0$. In this limit the amplitude diverges. The higher-order catastrophes diverge with higher fractional powers of $\hbar^{-1}$. Thus, the torus caustics mark the semiclassical function in a characteristic way analysed by Berry, Hannay and Ozorio de Almeida (1983). We can compare (7.43) with its crude semiclassical approximation (7.24) by means of the asymptotic formula

$$Ai(x) \xrightarrow[x \to \infty]{} \pi^{-1/2}(-x)^{-1/4} \cos\{\tfrac{2}{3}(-x)^{3/2} - \pi/4\}, \tag{7.44}$$

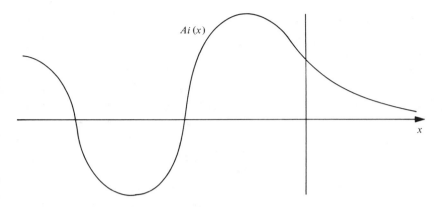

Fig. 7.2. Graph of the Airy function. The origin separates the region of exponential decay from the slowly decreasing oscillations.

given by Abramowitz and Stegun (1964, p. 448). Substituting this into (7.43) and using (7.42), we obtain

$$\langle q|\psi_I \rangle = c2^{1/2}\{\exp[i\hbar^{-1}S_+(q,I) - i\pi/4] + \exp[i\hbar^{-1}S_-(q,I) + i\pi/4]\}.$$
(7.45)

So we recover the semiclassical approximation, which is valid away from the caustic. As a bonus we have deduced that the phase change is $\pi/2$ for crossing the fold from the branch S_+ to S_-; that is, we have determined the phases α_j in (7.24).

The wave function (7.43) corresponds exactly to the momentum wave function $\exp\{i\hbar^{-1}[(p^3/3) - pI]\}$. Generally a semiclassical momentum wave function near a fold can be brought to this form by a canonical transformation of the phase space coordinates. But $S(p, I)$ will usually have a more complicated dependence on I, so the amplitude of the wave function will not be constant, and it will also have to be multiplied by the Jacobian of the transformation. This considerably complicates the task of deriving a *uniform approximation* to $\langle q|\psi \rangle$ based on (7.40) or (7.38). The problem is to guarantee that each stationary point contributes with its correct amplitude. The general method presented by Berry (1976) provides excellent quantitative results. The qualitative features are analogous to our simple example.

7.3 The energy spectrum

We found in the previous section that the semiclassical wave function gains a phase of $\pi/2$ as it passes from a branch where $\partial q_1/\partial p_1 < 0$ to one where

$\partial q_1/\partial p_1 > 0$. The equation that determines a fold is just $dq_1/dp_1 = 0$ for a one-freedom system, whereas if $L > 1$, we usually have to transform to new coordinates $(p_1, \mathbf{P}, q_1, \mathbf{Q})$, such that $\partial q_1/\partial p_1 = 0$ is the fold equation. The *Maslov index* μ for a closed curve is the number of times that $\partial q_1/\partial p_1$ becomes positive minus the number of times that $\partial q_1/\partial p_1$ becomes negative, as we go once around the curve. The definition is the same for any number of freedoms. If $L > 1$, then the curve is on a Lagrangian surface (manifold) and $\partial q_1/\partial p_1$ changes sign each time the curve crosses a fold, Arnold (1967) proved that the Maslov index is a topological invariant of a Lagrangian torus.

For an anticlockwise circuit along a closed curve on the (p, q) plane, such as fig. 7.3, the Maslov index is $\mu = 2$, so the wave function gains the total phase $\mu\pi/2 = \pi$ on returning to the point q_0 after a whole turn. Let us follow through the changes of the wave function step by step. If we define $S_+(q)$ such that $S_+(q_0) = 0$, then the wave function for the upper branch is

$$\langle q|\psi_+ \rangle = A_+(q)\exp\{i\hbar^{-1}S_+(q, I)\} \qquad (7.46)$$

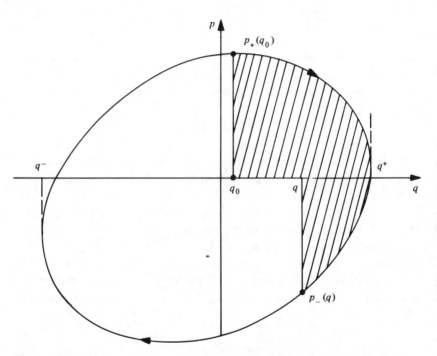

Fig. 7.3. The phase of the wave function, as we proceed clockwise, depends on the shaded area and on the phase gains of $\pi/2$ as a caustic is traversed. Quantization is obtained by requiring self-consistency of the wave function.

within a constant phase. Beyond the fold at q^+, we have

$$\langle q | \psi_- \rangle = A_-(q) \exp\{i\hbar^{-1} S_-(q, I) + i\pi/2\}, \qquad (7.47)$$

where

$$S_-(q, I) = \int_{q_0}^{q^+} p_+(q)\, dq + \int_{q^+}^{q} p_-(q)\, dq \qquad (7.48)$$

is the shaded area in fig. 7.3. On passing the other fold at q^-, we obtain

$$\langle q | \psi'_+ \rangle = A_+(q) \exp\{i\hbar^{-1} S'_+(q, I) + i\pi\}, \qquad (7.49)$$

where

$$S'_+(q, I) = S_+(q, I) + \oint p\, dq. \qquad (7.50)$$

The wave function cannot depend on the number of times we go around the closed curve. So the identity $\langle q | \psi'_+ \rangle = \langle q | \psi_+ \rangle$ determines the semiclassical quantization condition

$$\oint p\, dq = 2\pi\hbar(n + \tfrac{1}{2}). \qquad (7.51)$$

The presence of extra folds along the closed curve does not alter the Maslov index, because extra folds are created in pairs making opposite contributions to the index. The general condition (for $L = 1$) is therefore

$$\oint p\, dq = 2\pi\hbar\left(n + \frac{\mu}{4}\right). \qquad (7.52)$$

For a nonlinear oscillator $\mu = 2$, whereas for a rotor $\mu = 0$. If $L > 0$, we can choose L independent irreducible circuits γ_l in a way that they cross only folds, without touching any higher caustic. The self-consistency of the wave function after returning along any of these circuits entails the L conditions for torus quantization:

$$\int_{\gamma_l} \mathbf{p} \cdot d\mathbf{q} = 2\pi I_l = 2\pi\hbar\left(n_l + \frac{\mu_l}{4}\right). \qquad (7.53)$$

The quantization conditions go back to the old quantum theory. They are the result of work by Bohr, Sommerfeld, Einstein, Brillouin, Keller and Maslov.

The energy spectrum for an integrable system is given explicitly by the equations

$$E_\mathbf{n} = H[\hbar(\mathbf{n} + \boldsymbol{\mu}/4)]; \qquad (7.54)$$

that is, the eigenenergies correspond to level surfaces of the Hamiltonian,

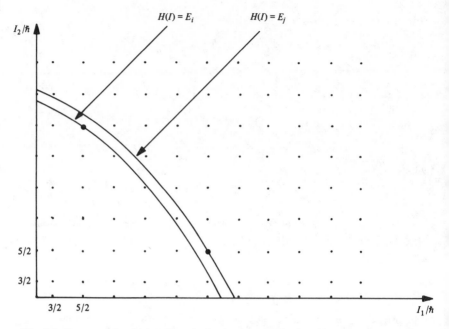

Fig. 7.4. In a semiclassical regime the lattice of quantized tori becomes quite dense. The sequence of quantized tori of increasing energy will not then generally be close together.

which touch some point of the lattice of quantized actions $I = \hbar(n + \mu/4)$. Figure 7.4 shows the case of a nonlinear oscillator $\mu = (2, 2)$.

7.4 The Weyl transform and the Wigner function

We shall be concerned mainly with the semiclassical limit of pure states. However, it should be noted that the torus quantization conditions imply that the average separation between energy levels tends to zero as $\hbar \to 0$. Hence, any imprecision in an energy measurement will prevent us from distinguishing the pure state of the system. In this case we must resort to a statistical description, by means of the *density operator*

$$\hat{\rho} = \sum_n a_n |\psi_n\rangle\langle\psi_n|. \tag{7.55}$$

In this formalism the probability of finding the system at the position $\mathbf{q}$ is given by the diagonal elements of the *density matrix* $\langle\mathbf{q}|\hat{\rho}|\mathbf{q}'\rangle$:

$$\langle\mathbf{q}|\hat{\rho}|\mathbf{q}\rangle = \sum_n a_n |\langle\mathbf{q}|\psi_n\rangle|^2. \tag{7.56}$$

Integrating over $\mathbf{q}$, we arrive at the interpretation of a_n as the probability that the pure state is $|\psi_n\rangle$. Thus, the density matrix for a pure state $|\psi_n\rangle$ has $a_m = \delta_{nm}$.

The *Weyl transform* of any operator $\hat{A}$ is defined as

$$A_w(\mathbf{p}, \mathbf{q}) \equiv \int d\mathbf{y} \langle \mathbf{q} + \mathbf{y}/2|\hat{A}|\mathbf{q} - \mathbf{y}/2\rangle \exp(-i\hbar^{-1}\mathbf{p}\cdot\mathbf{y}). \quad (7.57)$$

In this way we associate a function in phase space to any operator $\hat{A}$. By inverting this Fourier transform we obtain a unique operator $\hat{A}$ specified by $A_w(\mathbf{p}, \mathbf{q})$. In some simple cases $A_w(\mathbf{p}, \mathbf{q})$ will equal the classical variable $A(\mathbf{p}, \mathbf{q})$.

Exercises

1. Show that if $\hat{A} = f(\hat{\mathbf{q}})$, then $A_w(\mathbf{p}, \mathbf{q}) = f(\mathbf{q})$.
2. Show that if $\hat{A} = f(\hat{\mathbf{p}})$, then $A_w(\mathbf{p}, \mathbf{q}) = f(\mathbf{p})$.

It follows from these two results that the Weyl transform of a Hamiltonian $\hat{\mathbf{p}}^2/2m + V(\hat{\mathbf{q}})$ equals the corresponding classical Hamiltonian.

The Weyl transform of the density operator $\hat{\rho}/(2\pi\hbar)^L$,

$$W(\mathbf{p}, \mathbf{q}) \equiv (2\pi\hbar)^{-L} \int d\mathbf{x} \langle \mathbf{q} + \mathbf{y}/2|\hat{\rho}|\mathbf{q} - \mathbf{y}/2\rangle \exp(-i\hbar^{-1}\mathbf{p}\cdot\mathbf{y}), \quad (7.58)$$

is known as the *Wigner function* (Wigner, 1932). It furnishes a 'quasi-probability density' in phase space, in the sense that its projection,

$$\int W(\mathbf{p}, \mathbf{q}) \, d\mathbf{p} = \langle \mathbf{q}|\hat{\rho}|\mathbf{q}\rangle, \quad (7.59)$$

is a true probability density, though $W(\mathbf{p}, \mathbf{q})$ may be negative in some regions. Integrating (7.59) over $\mathbf{q}$ results in the normalization condition

$$\int W(\mathbf{p}, \mathbf{q}) \, d\mathbf{p} \, d\mathbf{q} = 1. \quad (7.60)$$

A pair of simple examples, discussed by Ozorio de Almeida and Hannay (1982), increases our intuitive grasp of the Wigner function for pure states. The first considers a travelling wave in the interval $(-d/2, d/2)$, with periodic boundary conditions:

$$\langle q|\psi\rangle = d^{-1/2} \exp(i\hbar^{-1}p_0 q), \quad (7.61)$$

where $p_0 = 2\pi n\hbar/d$. Inserting the corresponding density matrix into (7.58), we obtain

$$W(p,q) = \frac{1}{\pi d} \frac{\sin[d\hbar^{-1}(p-p_0)]}{p-p_0} \xrightarrow[\hbar \to 0]{} \frac{1}{d}\delta(p-p_0). \qquad (7.62)$$

The Wigner function is independent of q. In the semiclassical limit it 'condenses' onto the classical Liouville probability density – a δ function on the horizontal line $p = p_0$.

Replacing the boundary condition by hard walls at $\pm d/2$ leads to the wave function

$$\langle q|\psi\rangle = (2/d)^{1/2}\cos(\hbar^{-1}p_0 q). \qquad (7.63)$$

The corresponding Wigner function is

$$W(p,q) = \frac{1}{2\pi d}\left\{ \frac{\sin[d\hbar^{-1}(p-p_0)]}{p-p_0} + \frac{\sin[d\hbar^{-1}(p+p_0)]}{p+p_0} \right.$$
$$\left. + \frac{2\cos(2\hbar^{-1}p_0 q)\sin(d\hbar^{-1}p)}{p} \right\}. \qquad (7.64)$$

In the limit $\hbar \to 0$ the two first terms 'condense' onto the classical manifold $p = \pm p_0$, in the same way as for the travelling wave. But now there is also a nonclassical term. This interference term oscillates *along* the q axis itself, and it is essential for the correct oscillating projection of $|\langle q|\psi\rangle|^2$ onto the q axis:

$$\int dp\, W(p,q) = d^{-1}[1+\cos(2\hbar^{-1}p_0 q)] = 2d^{-1}\cos^2(\hbar^{-1}p_0 q). \quad (7.65)$$

The Wigner functions for both examples are depicted in fig. 7.5. In the classical limit, the oscillations of $|\langle q|\psi\rangle|^2$ become infinitely rapid, so we can measure only its average $\overline{|\langle q|\psi\rangle|^2} = 1/d$. By the same token, the interference term of the Wigner function can be considered to vanish as a consequence of its infinitely rapid positive and negative oscillations. The resulting Wigner function thus corresponds to the classical probability density.

An alternative representation of the Wigner function is

$$W(\mathbf{p},\mathbf{q}) = (2\pi\hbar)^{-L}\sum_n a_n\langle\psi_n|\hat{W}|\psi_n\rangle, \qquad (7.66)$$

introducing the *Weyl operator*,

$$\hat{W} = \int d\mathbf{y}|\mathbf{q}-\mathbf{y}/2\rangle\langle\mathbf{q}-\mathbf{y}/2|\exp(-i\hbar^{-1}\mathbf{p}\cdot\mathbf{y})$$

$$= \int d\mathbf{y}|\mathbf{p}-\mathbf{y}/2\rangle\langle\mathbf{p}+\mathbf{y}/2|\exp(+i\hbar^{-1}\mathbf{q}\cdot\mathbf{y}). \qquad (7.67)$$

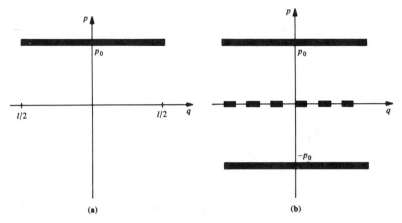

Fig. 7.5. Wigner function for a box with periodic conditions (a) and with hard walls (b).

From the Heisenberg equation for the evolution of an operator, we then obtain the time evolution of the Wigner function as

$$\frac{\partial W}{\partial t} = \frac{1}{i\hbar} \sum_n a_n \langle \psi_n | (\hat{W}\hat{H} - \hat{H}\hat{W}) | \psi_n \rangle, \qquad (7.68)$$

where the states $|\psi_n\rangle$ remain constant in time.

This equation takes on a particularly simple form in the case where $\hat{H}$ is a quadratic Hamiltonian, that is, for one degree of freedom:

$$2\hat{H} = H_{qq}\hat{q}^2 + H_{pq}(\hat{q}\hat{p} + \hat{p}\hat{q}) + H_{pp}\hat{p}^2. \qquad (7.69)$$

Exercise

Use (7.67) to demonstrate the relations

$$\hat{q}\hat{W} = \left(q + \frac{\hbar}{2i}\frac{\partial}{\partial p} \right)\hat{W}, \qquad \hat{W}\hat{q} = \left(q - \frac{\hbar}{2i}\frac{\partial}{\partial p} \right)\hat{W};$$
$$\qquad (7.70)$$
$$\hat{p}\hat{W} = \left(p - \frac{\hbar}{2i}\frac{\partial}{\partial q} \right)\hat{W}, \qquad \hat{W}\hat{p} = \left(p + \frac{\hbar}{2i}\frac{\partial}{\partial q} \right)\hat{W}.$$

Inserting these relations in (7.68), we obtain

$$\frac{\partial W}{\partial t} = \frac{1}{i\hbar} \sum_n a_n \langle \psi_n | \left[\frac{-\hbar}{i} H_{qq} q \frac{\partial \hat{W}}{\partial p} + \frac{\hbar}{i} H_{pq} \left(-p\frac{\partial \hat{W}}{\partial p} + q\frac{\partial \hat{W}}{\partial q} \right) \right.$$
$$\left. + \frac{\hbar}{i} H_{pp} p \frac{\partial \hat{W}}{\partial q} \right] | \psi_n \rangle$$

$$= (H_{qq}q + H_{pq}p)\frac{\partial W}{\partial p} - (H_{pq}q + H_{pp}p)\frac{\partial W}{\partial q}$$

$$= \frac{\partial H}{\partial q}\frac{\partial W}{\partial p} - \frac{\partial H}{\partial p}\frac{\partial W}{\partial q} = \{H, W\}. \tag{7.71}$$

Thus, the evolution of the Wigner function in this case is exactly the same as that of the classical Liouville distribution, since (7.71) is just the continuity equation for the 'incompressible' flow in phase space. This result cannot be generalized for nonquadratic Hamiltonians. For instance, equation (7.7) establishes a quantum mechanical equivalence for nonlinear shear transformations in phase space. These do not alter the projections onto the $\mathbf{q}$ axes of the Wigner function, though the latter is not itself invariant.

A classical quadratic Hamiltonian generates a linear vector field. As we found in section 1.2, the corresponding flow constitutes a group of symplectic (linear canonical) transformations. The Wigner function is therefore invariant with respect to the symplectic transformations. Since the Wigner function is a complete representation of the quantum state, we can obtain a new density matrix $\rho(\mathbf{q}, \mathbf{q}')$ by taking the Weyl transform of $\rho(\mathbf{q}, \mathbf{q}')$, then performing the symplectic transformation $(\mathbf{p}, \mathbf{q}) \to (\mathbf{p}, \mathbf{q})$ and finally taking the inverse of the Weyl transform. Once a pure Wigner function has been derived, we obtain a new wave intensity $|\langle \mathbf{q}|\psi \rangle|^2$ simply by projecting $W(\mathbf{p}, \mathbf{q})$ onto the $\mathbf{q}$ axes. These can be chosen to lie on any Lagrangian plane.

7.5 Semiclassical Wigner functions

To obtain the semiclassical Wigner function for a pure state $|\psi_n\rangle$, we construct the density matrix $\langle \mathbf{q}|\psi_n\rangle\langle\psi_n|\mathbf{q}'\rangle$ with the semiclassical wave function (7.24) and insert it into (7.58). For one branch of the wave function corresponding to the torus $\mathbf{I}$, we then have

$$W_{\mathbf{I}}(\mathbf{p}, \mathbf{q}) = c^2(2\pi\hbar)^{-L}\int dy \exp\{i\hbar^{-1}[S(\mathbf{q} + \mathbf{y}/2, \mathbf{I}) - S(\mathbf{q} - \mathbf{y}/2, \mathbf{I}) - \mathbf{p}\cdot\mathbf{y}]\}$$

$$\times |\det\frac{\partial^2 S}{\partial\mathbf{q}\partial\mathbf{I}}(\mathbf{q} + \mathbf{y}/2, \mathbf{I})\det\frac{\partial^2 S}{\partial\mathbf{q}\partial\mathbf{I}}(\mathbf{q} - \mathbf{y}/2, \mathbf{I})|^{1/2}. \tag{7.72}$$

We expect that, in the classical limit, $W_{\mathbf{I}}$ will be small unless $\mathbf{p} = \mathbf{p}(\mathbf{q}) = \partial S/\partial\mathbf{q}$. Given this momentum, the stationary-phase condition for the rapid oscillations of the exponential term is $\mathbf{y} = 0$. So, following Berry (1977a), we gain a first (classical) approximation of the Wigner function by expanding the exponent to first order in **y**, while taking **y** = 0 in the amplitude:

$$W_1(\mathbf{p}, \mathbf{q}) = c^2(2\pi\hbar)^{-L}|\det \partial \mathbf{p}/\partial \mathbf{I}| \int d\mathbf{y} \exp\{i\hbar^{-1}[\mathbf{p}(\mathbf{q}, \mathbf{I}) - \mathbf{p}]\cdot\mathbf{y}\}$$

$$= c^2|\det \partial \mathbf{p}/\partial \mathbf{I}|\delta[\mathbf{p} - \mathbf{p}(\mathbf{q}, \mathbf{I})]. \tag{7.73}$$

With the change of variable from **p** to **I**, we thus attain the symmetric form

$$W_1(\mathbf{p}, \mathbf{q}) = (2\pi)^{-L}\delta(\mathbf{I}(\mathbf{p}, \mathbf{q}) - \mathbf{I}), \tag{7.74}$$

normalized according to (7.60).

The classical limit of the pure-state Wigner function is a δ function over the corresponding torus. The amplitude of the δ function is constant in the conjugate variables $\boldsymbol{\phi}$, which do not generally coincide with Euclidean lengths on the torus. Consider, for instance, the one-dimensional tori in figs. 3.1a and b. The motion on the tori that nearly touch an unstable equilibrium is very slow in its vicinity. Most of the range of $0 \le \phi < 2\pi$ thus lies very close to this point. In consequence, the amplitude of the Wigner function is concentrated in the neighbourhood of the unstable equilibrium. In the case of a separable system with two freedoms, for which figs. 3.1a and b are sections, the equilibrium represents in fact an unstable periodic orbit. The Wigner function is concentrated near the periodic orbit, which is said to *scar* it. Projecting the Wigner function, we obtain a scar on the wave intensity near the periodic orbit. This coincides with the caustic in the case of a projection onto the q axis, but in the p representation the scar is far removed from any caustic. The wave function is then unusually concentrated near the unstable periodic orbit, even though its semiclassical quantization condition still depends on the action for the entire torus.

Given a mixed state (7.55), there will be a superposition of the δ functions for each of the quantized tori. But in the classical limit, the mean spacing between the quantized tori tends to zero. If the probability for the nth state has the limiting classical form

$$a_{\mathbf{n}} = a[(\mathbf{n} + \boldsymbol{\alpha}/4)\hbar], \tag{7.75}$$

then the Wigner function

$$W(\mathbf{p}, \mathbf{q}) = \sum_{\mathbf{n}} a_{\mathbf{n}}\delta\left(\mathbf{I}(\mathbf{p}, q) - \left(\mathbf{n} + \frac{\boldsymbol{\alpha}}{4}\right)\hbar\right) \xrightarrow[\hbar\to 0]{} a[\mathbf{I}(\mathbf{p}, \mathbf{q})], \tag{7.76}$$

since we can approximate the sum by an integral over **I**.

Without presuming that the pure Wigner is appreciable only when close to the torus, we obtain its semiclassical approximation by means of the stationary-phase approximation. The phase in (7.72) will be stationary for any **y** satisfying

$$\frac{\partial S}{\partial \mathbf{y}}\left(\mathbf{q}+\frac{\mathbf{y}}{2}\right)-\frac{\partial S}{\partial \mathbf{y}}\left(\mathbf{q}-\frac{\mathbf{y}}{2}\right)=\mathbf{p}\left(\mathbf{q}-\frac{\mathbf{y}}{2}\right)+\mathbf{p}\left(\mathbf{q}-\frac{\mathbf{y}}{2}\right)=2\mathbf{p}. \qquad (7.77)$$

These conditions define the end points of *Berry chords*, that is, the chords of the torus that have $(\mathbf{p}, \mathbf{q})$ as their midpoint. Their geometry is discussed by Berry (1977a) and Ozorio de Almeida and Hannay (1982). Figure 7.6 shows a Berry chord for a one-dimensional torus. Given a solution $\mathbf{y}_0$ of (7.77), the stationary phase

$$\hbar^{-1}\Phi_0 = \hbar^{-1}[S(\mathbf{q}+\mathbf{y}_0/2, \mathbf{I})-S(\mathbf{q}-\mathbf{y}_0/2, \mathbf{I})-\mathbf{p}\cdot\mathbf{y}_0] \qquad (7.78)$$

coincides precisely with the simplectic area (in units of $\hbar$) bounded by the chord and any path in the torus joining the end points. There is a pair of stationary points $\mathbf{y}_j$ and $-\mathbf{y}_j$ for the jth Berry chord, so the semiclassical

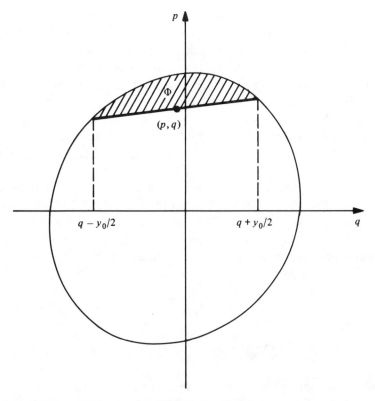

Fig. 7.6. A Berry chord for a one-dimensional torus. The symplectic area between the chord and the torus determines the phase of the Wigner function.

Wigner function is

$$W_1(\mathbf{p}, \mathbf{q}) = 2(\pi(2\pi\hbar)^{1/2})^{-L} \sum_j V_j^{-1/2} \cos\left(\hbar^{-1}\Phi_j - n_j\frac{\pi}{4}\right). \qquad (7.79)$$

The reader will find detailed specifications of the phase corrections $n_j\pi/4$, and the *amplitudes* $V_j^{-1/2}$ in Berry (1977a) and Ozorio de Almeida and Hannay (1982). The important point is that $V_j \to 0$ as the evaluation point $(\mathbf{p}, \mathbf{q})$ tends to the torus. Indeed, the torus belongs to a hypersurface of $2L - 1$ dimensions, the *Wigner caustic*, where the semiclassical approximation to the Wigner function diverges. Inside the Wigner caustic the Wigner function oscillates rapidly according to (7.79), while it vanishes outside. These singularities of the semiclassical Wigner function are a shortcoming of the stationary-phase method used in its derivation. At least one Berry chord shrinks as the evaluation point approaches the torus; that is, a pair of stationary points coalesces. Berry (1977a) and Ozorio de Almeida (1983) derive improved uniform approximations (based on diffraction catastrophe integrals), which remain finite on the Wigner caustic for $\hbar > 0$. The semiclassical equivalence of the Wigner representation with the semi-classical wave function of section 7.2 was obtained by Berry and Wright (1980) on the basis of remarkable projection identities.

The singularity of the semiclassical Wigner function at the Wigner caustic mimics a true peak of its amplitude. The difference between the torus and the rest of the Wigner caustic is that the phase remains constant on the entire torus, whereas it oscillates rapidly *along* the Wigner caustic as a whole. The classical limit (7.74) must therefore be understood in the same manner as the second example in the previous section – it does not involve the vanishing of the amplitude of the Wigner function, but the cancellation through arbitrarily rapid phase oscillations of the Wigner function, everywhere except on the torus itself.

We derived the semiclassical Wigner function from the wave function corresponding to a single branch of the torus. However, the symplectic invariance of the Wigner function implies that the result is quite general. If a Berry chord does not fit into a given branch of the torus, we can find a coordinate transformation fitting it in a single branch, as in fig. 7.7. The only chords for which this expedient is of no avail have both tips on branch boundaries for some representation: Their midpoints lie on the Wigner caustic. In any case, the symplectic invariance of (7.79) is evident, since V_j and n_j are canonically invariant. The transformation being linear, it takes a Berry chord into a new Berry chord with a corresponding midpoint. The symplectic area Φ is also preserved, because the transformation is canonical. Conversely, the noninvariance of the Wigner function with respect to general canonical transformations manifests itself clearly in the semiclassical limit: The Wigner function for the transformed torus depends on its new Berry chords and not on the curvilinear images of the old chords.

An interesting case is that of the transformation to the systems' own action-angle variables. In these coordinates a torus in the phase plane becomes a straight line, whereas the image of the Berry chord is now a curve. However, the semiclassical Wigner function in the action-angle variables is a discrete, evenly spaced set of δ functions in I with the separation $\hbar/2$. In the semiclassical limit we can construct the Wigner function in any convenient canonical coordinates, even if there is no exact quantum mechanical unitary transformation corresponding to this change of representation.

It should be noted that the Wigner function represents a single state, so it cannot supply directly matrix elements between different states (i.e., transition probabilities, etc.). These are obtained from the Moyal matrix or crossed Wigner function introduced by Moyal (1949). The semiclassical limit of the Moyal matrix for classically integrable systems in the variables $(\mathbf{p}, \mathbf{q})$ and $(\mathbf{I}, \boldsymbol{\phi})$ is discussed in Ozorio de Almeida (1984).

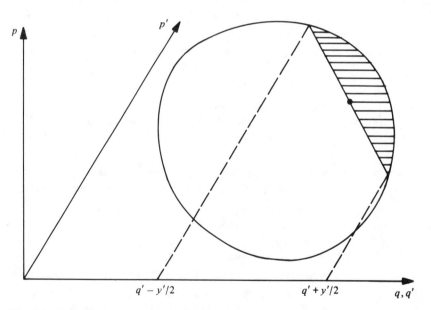

Fig. 7.7. If the centre of a Berry chord does not lie on the Wigner caustic, we can find a symplectic transformation such that the chord lies in a single branch of the action function.

7.6 The semiclassical propagator

In the beginning of this chapter, we found that the propagation of nonstationary wave functions, corresponding to general Lagrangian surfaces that move under the action of the classical Hamiltonian, are not constrained like stationary states to the invariant tori of integrable systems. The time-dependent semiclassical wave functions are formally impervious to the nonintegrability of the system, though the details of their evolution are certainly sensitive to the classical dynamics, as shown by Berry and Balazs (1979). One of the methods for extracting information about stationary states of nonintegrable systems, to be discussed in chapter 9, is to consider the way the Hamiltonian propagates nonstationary states.

It is particularly convenient to define states related to the *evolution operator*

$$\hat{U}(t) = \exp(-i\hbar^{-1}\hat{H}t). \tag{7.80}$$

Its coordinate representation $\langle q | \hat{U} | q \rangle$, known as the *propagator*, can be identified as the wave function $\langle q | \psi(t) \rangle$ that results from the propagation

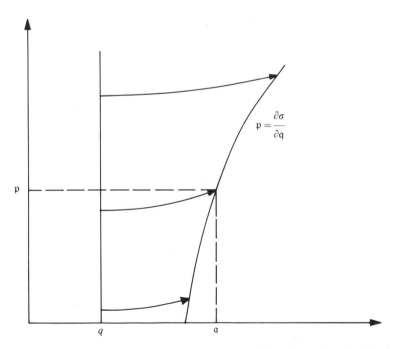

Fig. 7.8. The propagator for a system with one freedom is associated with the time evolution of a vertical $q = $ const. line.

of $\langle \mathbf{q} | \mathbf{q} \rangle = \delta(\mathbf{q} - \mathbf{q})$ in the time t. Classically this initial state corresponds to a uniform $\mathbf{p}$ distribution along the plane $\mathbf{q} = \mathbf{q}$. This plane will evolve into the Lagrangian surface

$$\mathbf{p} = \frac{\partial \sigma}{\partial \mathbf{q}}(\mathbf{q}, \mathbf{q}, t), \tag{7.81}$$

where σ is a solution of the Hamilton–Jacobi equation (7.10). The configuration for a system with one freedom is sketched in fig. 7.8.

The semiclassical propagator has the standard form (7.23), except for the presence of caustics. Of course, the initial wave function $\langle \mathbf{q} | \mathbf{q} \rangle$ is a single degenerate caustic! So we resort to the Maslov method, since the $\mathbf{p}$ representation

$$\langle \mathbf{p} | \hat{U}(t) | \mathbf{q} \rangle = (2\pi\hbar)^{-1/2} \left| \det \frac{\partial^2 \sigma}{\partial \mathbf{p} \, \partial \mathbf{q}} \right|^{1/2} \exp\{i\hbar^{-1}\sigma(\mathbf{p}, \mathbf{q}, t)\} \tag{7.82}$$

is guaranteed to have no caustics for small enough t.

Exercise

Show that for small enough positive time the semiclassical propagator has the form

$$\langle \mathbf{q} | \hat{U}(t) | \mathbf{q} \rangle = (2\pi i\hbar)^{-1/2} \left| \det \frac{\partial^2 \sigma}{\partial \mathbf{q} \, \partial \mathbf{q}} \right|^{1/2} \exp\{i\hbar^{-1}\sigma(\mathbf{q}, \mathbf{q}, t)\} \tag{7.83}$$

if $H = \mathbf{p}^2/2m + V(\mathbf{q})$.

For $t < 0$ the unitary property of the evolution operator implies that

$$\langle \mathbf{q} | \hat{U}(-t) | \mathbf{q} \rangle = \langle \mathbf{q} | \hat{U}(t) | \mathbf{q} \rangle^*. \tag{7.84}$$

Even for simple motions where $\dot{\mathbf{q}} = \mathbf{p}$, the evolving Lagrangian surface may eventually develop caustics. In this instance they are known as *focal points* or *conjugate points* to $\mathbf{q}$ – they are the loci beyond which the $\mathbf{q}$ orbits emanating from the point $\mathbf{q}$ touch again, as at the cusp in fig. 7.9. The Morse index μ for a $\mathbf{q}$ orbit is the number of times that it crosses a caustic. Arnold (1967) established the connection between the Morse index and the Maslov index and hence the canonical invariance of the former. We know that the presence of caustics on a Lagrangian surface leads to phase jumps for the wave functions corresponding to each of its branches. Fortunately, the phase for the jth branch can be obtained from the Morse index μ_j of the corresponding $\mathbf{q}$ orbits. Except for the neighbourhood of

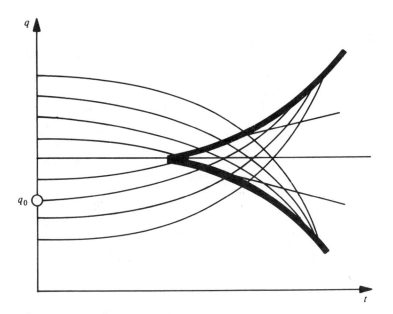

Fig. 7.9. The Morse index of the orbit starting at q_0 changes from 0 to 1 as it enters the region within the cusp.

caustics, the full semiclassical propagator is

$$\langle \mathbf{q} | \hat{U}(t) | \mathbf{q} \rangle = (2\pi i \hbar)^{-1/2} \sum_j \left| \det \frac{\partial^2 \sigma_j}{\partial \mathbf{q} \, \partial \mathbf{q}} \right|^{1/2} \exp \left\{ i\hbar^{-1} \sigma_j(\mathbf{q}, \mathbf{q}, t) - i\mu_j \frac{\pi}{2} \right\}.$$

(7.85)

The derivation of the role of the Morse index in (7.85) is simplest in the Feynman path integral formalism used by Gutzwiller (1967) and by Berry and Mount (1972).

The propagator is the coordinate representation of the unitary evolution operator. The unitary transformation for a fixed t corresponds classically to the canonical transformation $(\mathbf{p}, \mathbf{q}) \rightarrow (\mathbf{p}, \mathbf{q})$, generated implicitly by $\sigma(\mathbf{q}, \mathbf{q}, t)$ according to

$$\mathbf{p} = \frac{\partial \sigma}{\partial \mathbf{q}} (\mathbf{q}, \mathbf{q}, t); \qquad -\mathbf{p} = \frac{\partial \sigma}{\partial \mathbf{q}} (\mathbf{q}, \mathbf{q}, t).$$

(7.86)

Since the coordinate representation of the evolution operator depends only on the final canonical transformation, we can associate an approximately unitary operator $\hat{U}$ to a canonical transformation generated by any arbitrary function $\sigma(\mathsf{q}, \mathbf{q})$.

If the transformation is symplectic, the vertical line $\mathsf{q} = q$ will be transformed into another straight line. So there will be no caustics, unless the straight line is vertical. The generating function must have the quadratic form

$$\sigma(\mathsf{q}, q) = (\alpha/2)\mathsf{q}^2 + \beta\mathsf{q}q + (\gamma/2)q^2, \qquad (7.87)$$

so that the amplitude in (7.83) is just $\beta^{1/2}$. We can verify directly that the semiclassical propagator (7.83) then satisfies

$$\int dq' \langle \mathsf{q} | \hat{U}^{-1} | q' \rangle \langle q' | \hat{U} | q \rangle = \int dq' \langle q' | \hat{U} | \mathsf{q} \rangle^* \langle q' | \hat{U} | q \rangle$$

$$= (2\pi\hbar i)^{-1} \exp\{i\hbar^{-1}\gamma(q^2 - \mathsf{q})^2/2\} \int dq' \, \beta \exp\{i\hbar^{-1}\beta(q - \mathsf{q})q'\}$$

$$= \delta(q - \mathsf{q}). \qquad (7.88)$$

Thus, the semiclassical form of the coordinate representation of the unitary operator corresponding to a symplectic transformation is exact.

8

Quantization of ergodic systems

The remarkable self-consistency of the semiclassical theory of integrable systems prevents its natural generalization in the absence of invariant tori. The two ways we may attempt to tackle the problem are by studying the quantization of quasi-integrable systems or, at the other extreme, by plunging straight into ergodic systems. The first option has so far been confined to the quantization of the nearest integrable system that can be found. If the perturbations from integrability are sufficiently weak, this procedure can be justified by the analysis presented in section 9.4.

The present chapter starts with the investigation of the Berry–Voros hypothesis that the classical limit of the Wigner function for an ergodic system is uniform on the energy shell. This leads to some interesting consequences for the smoothed probability density and spatial correlation function. The classical limit of the Wigner function can be refined through multiple iterations of the pure-state condition. In section 8.2 it is shown that the torus Wigner function satisfies the pure-state condition, whereas the energy shell Wigner function is brought into a form that depends on individual orbits. A very important feature is that for an ergodic system we can identify the Wigner caustic with the energy shell itself – the semiclassical Wigner function oscillates inside and decays outside the shell.

The 'Weyl rule' for the mean density of eigenenergies, derived in section 8.3, is insensitive to the integrable or chaotic nature of the classical motion. However, the fluctuations of the energy spectrum of some ergodic systems have been computationally verified to have an 'ergodic property' – their correlation functions and averages coincide with those of an ensemble of random matrices. A brief introduction to random matrix theory is given in section 8.4.

Finally, we return once again to linear maps on the torus. They can be quantized exactly, though at the cost of discretizing the phase space. It turns out that they break the rule that all classically ergodic systems have an ergodic energy spectrum.

8.1 A Wigner function on the energy shell

The Wigner function corresponding to a quantized torus was shown in section 7.5 to condense onto the torus as a δ function in the classical limit. This result is intuitively satisfying: Almost all orbits of an integrable system wind uniformly around some torus. The classical probability density of finding the system anywhere on the torus is therefore uniform, and there is no possibility of finding it off the torus.

The situation for an ergodic system is analogous: Almost all orbits wind uniformly over the energy shell. The *Berry–Voros hypothesis* (Berry, 1977b; Voros, 1979) is that the Wigner function for the stationary state of an ergodic system is then approximately

$$W_c(\mathbf{p}, \mathbf{q}) = C \,\delta\{E - H(\mathbf{p}, \mathbf{q})\}, \tag{8.1}$$

where the normalization constant is

$$C^{-1} = \int d\mathbf{p}\, d\mathbf{q}\, \delta\{E - H(\mathbf{p}, \mathbf{q})\}. \tag{8.2}$$

For an integrable system we know that the correct semiclassical approximation (7.79) contains interference terms neglected in the classical approximation (7.74). For an ergodic system it is hard to go beyond (8.2), though a method to achieve this will be put forward in the next section.

Let us consider the projection $|\langle \mathbf{q}|\psi\rangle|_m^2$ of the classical approximation of a Wigner function onto configuration space. In the case of a torus we have

$$|\langle \mathbf{q}|\psi\rangle|_m^2 = \int d\mathbf{p}\, W_c(\mathbf{p}, \mathbf{q}) = (2\pi)^{-L} \sum_i \left| \det \frac{\partial \boldsymbol{\phi}}{\partial \mathbf{q}}(\mathbf{p}_i(\mathbf{q}), \mathbf{q}) \right|, \tag{8.3}$$

where the $\boldsymbol{\phi}$ are the angle variables on the torus. Comparing this with the square of (7.24), we find that the contribution of each layer of the torus is correct, though (8.3) ignores the interference between layers. These interference terms may be neglected as $\hbar \to 0$, but we also obtain this limit by averaging the semiclassical intensity over a small region D^L, large enough to contain many oscillations of the wave function:

$$|\langle \mathbf{q}|\psi\rangle|_m^2 = D^{-L} \int_{D/2}^{D/2} \cdots \int_{D/2}^{D/2} d\mathbf{y}\, |\langle \mathbf{q} + \mathbf{y}|\psi\rangle|^2. \tag{8.4}$$

By analogy we also interpret the projections of the energy shell Wigner function (8.2) as averages over the unknown semiclassical wave intensity. In the case of the Hamiltonian

$$H(\mathbf{p}, \mathbf{q}) = p^2/2m + V(\mathbf{q}), \tag{8.5}$$

the averaged intensity (8.4) will be

$$|\langle \mathbf{q}|\psi\rangle|_m^2 = \frac{\int dp\, p^{L-1}\, \delta(E - p^2/2m + V(\mathbf{q}))}{\int d\mathbf{q}\, dp\, p^{L-1}\, \delta(E - p^2/2m + V(\mathbf{q}))}$$

$$= \frac{(E - V(\mathbf{q}))^{L/2-1}\,\Theta(E - V(\mathbf{q}))}{\int d\mathbf{q}(E - V(\mathbf{q}))^{L/2-1}\,\Theta(E - V(\mathbf{q}))}, \qquad (8.6)$$

where $\Theta(x)$ is the unit step function. The classical probability density is nonzero only if $V(\mathbf{q}) < E$.

Let us consider the boundary of this region. For $L = 1$ the averaged intensity is singular at the boundary point. This is just what we expect – after all, the system is integrable, so we are dealing with a caustic. For $L = 2$ the density is constant, so the classically allowed region terminates with a finite step of the intensity. For $L > 2$ the probability density goes smoothly to zero. This behaviour for $L \geq 2$ contrasts strongly with that of integrable systems. These boundaries are therefore named *anticaustics* by Berry (1977b). We have seen that the caustics of integrable systems do not correspond to true singularities of the wave intensity. However, as $\hbar \to 0$ we do obtain sharp ridges near the caustic. The Berry–Voros hypothesis predicts the absence of this important feature in the intensity of the states of ergodic systems.

The Wigner function contains more information than its projection. In particular we obtain the *correlation function* $C(\mathbf{y}; \mathbf{q})$ by averaging the variable $\mathbf{q}$ in the density matrix $\langle \mathbf{q} + \mathbf{y}/2|\psi\rangle\langle\psi|\mathbf{q} - \mathbf{y}/2\rangle$ over a small region D^L, while keeping $\mathbf{y}$ fixed:

$$C(\mathbf{y}; \mathbf{q}) = \frac{\{\langle \mathbf{q} + \mathbf{y}/2|\psi\rangle\langle\psi|\mathbf{q} - \mathbf{y}/2\rangle\}_m}{|\langle \mathbf{q}|\psi\rangle|_m^2}$$

$$= |\langle \mathbf{q}|\psi\rangle|_m^{-2} \int d\mathbf{p}\, W_c(\mathbf{p}, \mathbf{q})\exp(i\hbar^{-1}\mathbf{p}\cdot\mathbf{y}). \qquad (8.7)$$

For the one-dimensional wave function (7.63), the density matrix is

$$\langle q + y/2|\psi\rangle\langle\psi|q - y/2\rangle = d^{-1}\{\cos(\hbar^{-1}p_0 y) + \cos(\hbar^{-1}2p_0 q)\}. \quad (8.8)$$

Averaging over q in a region large enough for the second term to have many oscillations, we obtain

$$C(y) = \cos(\hbar^{-1}p_0 y), \qquad (8.9)$$

which coincides with the result of inserting the classical Wigner function

$$W_c(p, q) = (2d)^{-1}[\delta(p - p_0) + \delta(p + p_0)] \qquad (8.10)$$

in (8.7).

For states corresponding to tori

$$C(\mathbf{y};\mathbf{q}) = \left\{ \sum_j \left| \det\frac{\partial\boldsymbol{\phi}}{\partial\mathbf{q}} \right| \right\}^{-1} \sum_i \left| \det\frac{\partial\boldsymbol{\phi}}{\partial\mathbf{q}} \right| \exp\{i\hbar^{-1}\mathbf{p}_j(\mathbf{q})\cdot\mathbf{y}\}, \qquad (8.11)$$

a weighted average of plane de Broglie waves, with wave vectors $\mathbf{p}_j(\mathbf{q})/\hbar$. The torus has a finite number of layers, so $C(\mathbf{y},\mathbf{q})$ is necessarily anisotropic in $\mathbf{y}$. The resulting interference fringes are illustrated in fig. 8.1 in the simple case of a two-dimensional torus with only two layers over the point $\mathbf{q}$. For an ergodic system there is a continuum of $\mathbf{p}$'s for each $\mathbf{q}$. In the case of the simple Hamiltonian (8.5) we obtain

$$C(\mathbf{y};\mathbf{q}) = \left\{ \int d\Omega \right\}^{-1} \int d\Omega \exp\{i\hbar^{-1}[2m(E - V(\mathbf{q}))]^{1/2}\mathbf{y}\cdot\boldsymbol{\Omega}\}, \qquad (8.12)$$

where Ω is the unit vector in the $\mathbf{p}$ direction and $d\Omega$ is the differential of the L-dimensional angle in $\mathbf{p}$. According to Berry (1977b) this can be integrated as

$$C(\mathbf{y};\mathbf{q}) = \Gamma\left(\frac{N}{2}\right) \frac{J_{N/2-1}\{\hbar^{-1}[2m(E - V(\mathbf{q}))]^{1/2}|\mathbf{y}|\}}{\{\hbar^{-1}[2m(E - V(\mathbf{q}))]^{1/2}|\mathbf{y}|\}}, \qquad (8.13)$$

where Γ is the Euler gamma function and J_ν is a Bessel function (both defined in Abramowitz and Stegun, 1964). Thus, the Berry–Voros approxi-

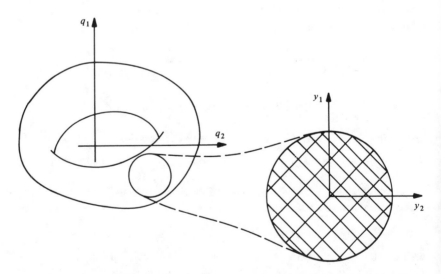

Fig. 8.1. The correlation function shows the interference fringes of a finite number of plane de Broglie waves.

mation to the correlation function of an ergodic system is isotropic in **y**. This prediction is confirmed by the calculations of McDonald and Kaufman (1979) for the stadium. They also discovered that the eigenfunctions exhibit a random pattern for the nodal curves.

8.2 The pure-state condition

The condition that the density operator $\hat{\rho}$, given by (7.55), defines a pure state is simply that

$$\hat{\rho}^2 = \hat{\rho}. \tag{8.14}$$

Taking the Weyl transform of both sides, we obtain the following identity between Wigner functions,

$$\int_{-\infty}^{\infty} d\mathbf{x}_1 \, d\mathbf{x}_2 \, W(\mathbf{x}_1) W(\mathbf{x}_2) \cos(\hbar^{-1}\Delta) = \left(\frac{\pi\hbar}{2}\right)^L W(\mathbf{x}), \tag{8.15}$$

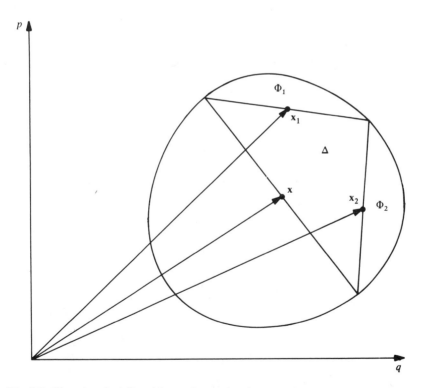

Fig. 8.2. The triangle Δ fits with two Berry chords to produce a third Berry chord in a stationary configuration.

where $x = (p, q)$ and Δ is twice the symplectic area of the parallelogram with sides $x_1 - x$ and $x_2 - x$, that is,

$$\Delta(x, x_1, x_2) = 2 \sum_{l=1}^{L} \det \begin{bmatrix} p_{1l} - p_l & p_{2l} - p_l \\ q_{1l} - q_l & q_{2l} - q_l \end{bmatrix}. \tag{8.16}$$

An alternative geometric interpretation of Δ is that of the symplectic area of the triangle that has x, x_1 and x_2 as the midpoints of its sides. As we see in fig. 8.2, this fits in beautifully with the semiclassical Wigner function, defined in terms of Berry chords. Indeed, for the case drawn, the phases of $W(x_1)$ and $W(x_2)$ together with Δ equal the phase of the Wigner function $W(x)$. Ozorio de Almeida (1983) showed that this is the configuration for the stationary-phase condition with respect to x_1 and x_2 in the integral (8.15). Thus, the semiclassical Wigner function satisfies the pure-state condition within the stationary-phase approximation.

The fact that condition (8.15) also holds for the amplitudes and phase corrections of the Wigner function (and that the results can be generalized to tori of any dimension) is just another example of the self-consistency of the semiclassical theory of integrable systems. The important question is whether we can use the pure-state condition to refine a crude approximation of the Wigner function.

Let us thus insert the classical approximation of the torus Wigner

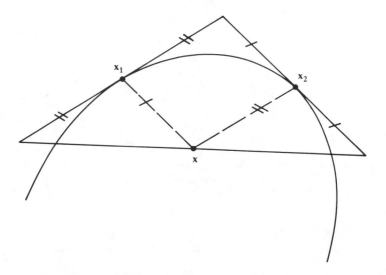

Fig. 8.3. Inserting classical Wigner functions in the pure-state condition, we obtain an approximation whose phase is that of a triangle tangent to the torus.

function into the integral in (8.15); there results a new approximation

$$W_1(\mathbf{x}) = (2\pi^3\hbar)^{-L} \int_{-\infty}^{\infty} d\mathbf{x}_1\, d\mathbf{x}_2 \delta(\mathbf{I}(\mathbf{x}_1) - \mathbf{I})\delta(\mathbf{I}(\mathbf{x}_2) - \mathbf{I}) \cos\{\hbar^{-1}\Delta(\mathbf{x}, \mathbf{x}_1, \mathbf{x}_2)\}.$$

$$(8.17)$$

For $L = 1$, it is easy to verify that the phase is stationary in the integral over the angle variables ϕ_1 and ϕ_2 when the triangle Δ is tangent to the torus at both points $\mathbf{x}_1$ and $\mathbf{x}_2$. For example, if we fix the vector $\mathbf{x}_2 - \mathbf{x}$ in fig. 8.3, the area of the parallelogram defined by this vector together with $\mathbf{x}_1 - \mathbf{x}$ is stationary with respect to small displacements of $\mathbf{x}_1$ along the torus. This result can also be generalized for $L > 1$: The phase $\hbar^{-1}\Delta$ is stationary when two sides of the triangle belong to tangent planes of the torus at $\mathbf{x}_1$ and $\mathbf{x}_2$. This can be seen by taking the tangent plane at $\mathbf{x}_1$ to be the $\mathbf{p} = 0$ coordinate plane (permitted by the Lagrangian property of the torus). The condition that the side of Δ at $\mathbf{x}_1$ is in the $\mathbf{p} = 0$ plane forces the projections of $(\mathbf{x}_2 - \mathbf{x})_l$ onto the L conjugate planes (p_l, q_l) to be parallel to the q_l axis (fig. 8.4). None of the areas A_l will be altered by displacements of $\mathbf{x}_1$ along the $\mathbf{p} = 0$ plane. So the phase is stationary with

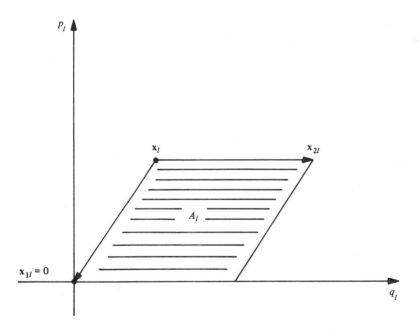

Fig. 8.4. The stationary-phase condition is that $(\mathbf{x}_2 - \mathbf{x})$ be parallel to the tangent plane to the torus at $\mathbf{x}_1$. This can be taken as the $\mathbf{p} = 0$ plane.

respect to x_1 if it lies on a tangent plane, the same condition holding for x_2. Note that the L-dimensional tangent planes at x_1 and x_2 intersect at a single point in the $2L$-dimensional phase space. This is a vertex of Δ, the phase of $W_1(x)$; that is, we obtain a clumsy imitation of the Berry chord construction. Thus, after a single iteration of the pure-state condition, we already have an approximation to the Wigner function that oscillates inside the torus, a part of the Wigner caustic.

Successive iterations of the pure-state conditions bring further refinements to the Wigner function. It is easily seen that the stationary-phase condition for $W_2(x)$, obtained by inserting $W_1(x_1)$ and $W_1(x_2)$ in (8.15), is that the pentagon drawn in fig. 8.5 has four sides tangent to the torus at their midpoints. Further iterations bring improved polygonal approximations to the Berry chord phase, which emerges in the limit of infinite iterations!

Let us now apply our method for refining approximations of the Wigner function to the Berry–Voros hypothesis. The first iteration of the pure-state condition is

$$W_1(x) = \frac{c^2}{(\pi\hbar/2)^L} \int dx_1\, dx_2\, \delta(H(x_1) - E)\delta(H(x_2) - E)\cos(\hbar^{-1}\Delta). \quad (8.18)$$

We seek the condition for the area Δ to be stationary with respect to the $(2L-1)$ free variables in x_1 and x_2. Again we choose the $(2L-1)$-dimensional plane tangent to the energy shell at x_1 as a coordinate plane

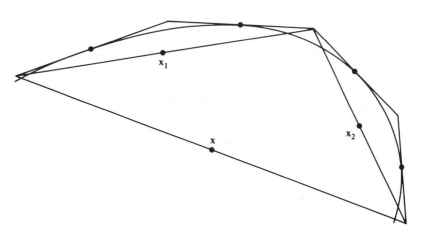

Fig. 8.5. The second iteration of the pure-state condition leads to an approximate Wigner function whose phase depends on the area of a tangent pentagon.

with $p_1 = 0$. If the projection of the vector $\mathbf{x}_2 - \mathbf{x}$ onto the plane (p_1, q_1) is parallel to the q_1 axis as in fig. 8.4, then the area A_1 will be stationary with respect to displacements of $\mathbf{x}_1$. However, having the side of Δ at $\mathbf{x}_1$ contained in the $p_1 = 0$ plane imposes no other restriction on $\mathbf{x}_2 - \mathbf{x}$, so the other areas A_j need not be stationary. The only way of guaranteeing that the full *symplectic area* Δ is stationary is for $\mathbf{x}_2 - \mathbf{x}$ to be parallel to the q_1 axis itself. In a Euclidean space the situation would be hopeless, because this direction would not be uniquely defined on the tangent plane. In a symplectic space the q_1 axis does have a unique direction, namely $\dot{\mathbf{x}} = \mathscr{J} \, \partial H/\partial \mathbf{x}$, where $\mathscr{J}$ is given by (1.37). Thus, the phase of the Wigner function W_1 is that of the triangle Δ, where $\mathbf{x}_1$ and $\mathbf{x}_2$ are tangent to *classical orbits* of the ergodic system.

At the W_1 level there is no restriction identifying the orbits at $\mathbf{x}_1$ and $\mathbf{x}_2$. However, on approaching the limit of infinite iterations there arises a polygonal line with infinitesimal sides, each one being tangent to an orbit. In the limit W_∞ we obtain a Berry chord on the shell with both edges touching the same orbit. The corresponding phase is the symplectic area (in units of $\hbar$) bounded by the orbit and the chord. Note that, in this case, we canot choose any path between the chord tips, since the energy shell is a not a Lagrangian surface.

Far from the energy shell the semiclassical Wigner function will be very complicated. Near the shell we can approximate its $(2L - 2)$-parameter family of orbits by plane curves. Each point within and near the shell lies on a two-dimensional plane containing a single orbit. In each plane we obtain a semiclassical Wigner function whose structure is identical to the one for a single freedom (simultaneously integrable and ergodic!). The energy shell is the Wigner caustic – a fold. Beyond it the Wigner function decays, whereas inside there are multiple-phase oscillations of a complex nature yet to be analysed. The ergodic Wigner function has some relation to that for an integrable system with $L \geq 2$. However, the Wigner caustic of the latter is a mere geometric construction based on the invariant torus, whereas in the former case the Wigner caustic is the energy shell itself.

It is hard to calculate the amplitudes and phase corrections of the semi-classical Wigner function directly from the iterations of the pure-state condition. The important and surprising point is that, from the refinements of the energy shell Wigner function, there emerges a semiclassical dependence on individual orbits, for all the complexity of their structure. In chapter 9 we shall use a formalism based on the propagator, which takes the orbits into account from the start, and thus derive some important results in an easier way than iterating the pure-state condition.

8.3 The mean level density

The energy spectrum of integrable systems was derived in section 7.3 from a detailed consideration of the semiclassical wave functions. It is now important to try to obtain information about the spectrum, without relying on the (unknown) characteristics of the eigenstates of unintegrable systems. We shall deal with the *density of states*

$$d(E) = \sum_{j=1}^{\infty} \delta(E - E_j) = \text{Tr }\delta(E - \hat{H}) \tag{8.19}$$

and its integral

$$N(E) = \sum_{j=1}^{\infty} \Theta(E - E_j) = \text{Tr }\Theta(E - \hat{H}), \tag{8.20}$$

the *cumulative density* or *mode number*.

Turning once again to the formalism of Weyl and Wigner, we note that inverting the Fourier transform (7.57) and integrating over $\mathbf{q}$, we obtain

$$\text{Tr }\hat{A} = (2\pi h)^{-L} \int d\mathbf{p}\, d\mathbf{q}\, A_w(\mathbf{p}, \mathbf{q}). \tag{8.21}$$

We found in section 7.4 that in some simple cases the Weyl transform of an operator coincides with its classical limit. According to the correspondence principle this equivalence holds in the semiclassical limit (also verified directly using semiclassical Wigner functions by Ozorio de Almeida, 1984). So following Berry (1983), we substitute (8.21) for the right side of (8.20) with $[\Theta(E - \hat{H})]_w \simeq \Theta(E - H(\mathbf{p}, \mathbf{q}))$, which results in the *Weyl rule*:

$$\bar{N}(E) = (2\pi\hbar)^{-L} \int d\mathbf{p}\, d\mathbf{q}\, \Theta(E - H(\mathbf{p}, \mathbf{q})). \tag{8.22}$$

The surprising feature of this formula is its complete generality. No distinction is made as to the dynamical nature of the system, whether integrable, ergodic or otherwise. The approximate number of states with energy smaller than E is determined by the volume within the corresponding energy shell, measured in units of $(2\pi\hbar)^L$.

This continuous density can only be an approximation of the steps in the mode number (8.20). That it represents an averaged density becomes clear from the consideration of integrable systems. For these the mode number, derived in section 7.3, is precisely

$$N(E) = \sum_{\mathbf{n}} \Theta(E - H(\mathbf{I_n})), \tag{8.23}$$

where the I_n are the actions of the quantized invariant tori. These form a regular lattice in the action variables, so as a first approximation we can take

$$\bar{N}(E) = \hbar^{-L} \int d\mathbf{I}\, \Theta(E - H(\mathbf{I})). \tag{8.24}$$

Since the Hamiltonian is independent of the angle variables, this integral can be rewritten as

$$\bar{N}(E) = \hbar^{-L} \int d\mathbf{I}\, \frac{d\boldsymbol{\theta}}{(2\pi)^L}\, \Theta(E - H(\mathbf{I})). \tag{8.25}$$

The canonical transformation $(\mathbf{I}, \boldsymbol{\theta}) \rightarrow (\mathbf{p}, \mathbf{q})$ preserves volume; therefore, the mean level density (8.25) can be identified with the Weyl rule.

The verification of the Weyl rule for nonintegrable systems is possible only through 'computational experiments' with simple systems. The most popular systems are *quantum billiards*, that is, the quantization of the two-dimensional classical billiards presented in section 3.5. Within the boundaries, the Schrödinger equation reduces to the Helmholtz equation, so the problem is simply to ascertain the vibrational spectra of drums with different shapes. The boundary conditions on the square in Sinai's billiard shown in fig. 3.12 are periodic. Berry (1981b) calculated the energy spectrum of Sinai's billiard for five values of the parameter R, the ratio of the radius of the circle to the side of the square. The results appear in fig. 8.6. It should be noted that the system is integrable for $R = 0$, whereas in all other cases it is ergodic with positive entropy.

For a particle with unit mass in a billiard, the Weyl rule takes the simple form

$$\bar{N}(E) \simeq AE/(2\pi\hbar^2), \tag{8.26}$$

where A is the available area of the billiard. Unfortunately, the discontinuity of the potential, which helps the computation of eigenvalues, hinders the agreement of the Weyl rule with the averaged mode number. Even so, Baltes and Hilf (1978) provide corrections to the Weyl rule, which also take into account the perimeter γ of the billiard, its curvature, its connectivity and the contribution of the vertices:

$$\bar{N}(E) = \frac{AE}{(2\pi\hbar)^2} - \frac{\gamma E^{1/2}}{2^{3/2}\pi\hbar} + K, \tag{8.27}$$

the latter contributions all being included in K, a constant in energy. This is the equation of the dashed curves in fig. 8.6, which correctly interpolate

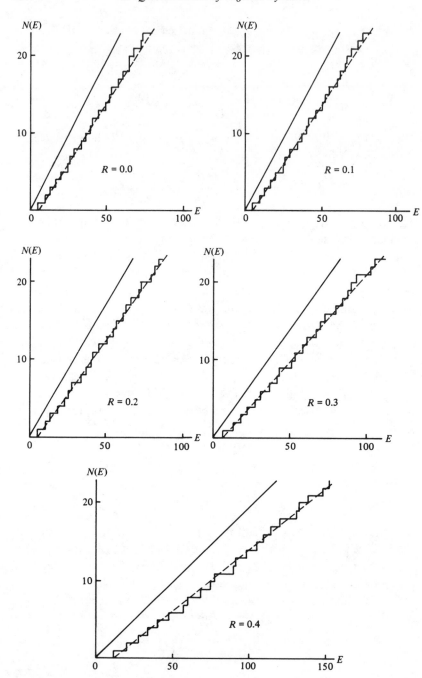

Fig. 8.6. The mode number for Sinai's billiard, calculated for five values of the parameter R. (Courtesy of M. V. Berry.)

the steps in the cumulative density of states. The Weyl rule (8.26) is presented as an unbroken straight line in each graph.

Even though it is necessary to add corrections to the Weyl rule in the case of billiards, the important fact is that we have a simple theory that predicts the correct mean level density. It is insensitive to the nature of the classical motion (also in the case of billiards), so that it is only more subtle properties of the spectrum – the fluctuations of the density about its mean – that may distinguish the dynamical nature of the corresponding classical motion. In the following section we shall briefly study the theory of random matrices and examine the evidence that they furnish correct predictions of the statistics of fluctuations for the spectra of classical chaotic systems. Chapter 9 provides the dynamical explanation of some of these results, based on the quantization of periodic orbits.

8.4 Random matrices

In the semiclassical limit the mean level density $\bar{d}(E)$ increases with $\hbar^{-L}$. For small $\hbar$ all the physically significant properties of the spectrum may depend only on $\bar{d}(E)$ and on some averages over the *fluctuations*

$$\tilde{d}(E) = d(E) - \bar{d}(E) \tag{8.28}$$

of the spectrum. These averages can then be defined locally for an energy range $\delta E \propto \hbar^{-1}$, containing $\delta N \propto \hbar^{-(L-1)}$. In other words, the semiclassical limit permits us to subject the spectrum to a statistical treatment even in an energy range that is classically small.

This section presents techniques for calculating statistical properties of spectra from those ensembles of random matrices that share only the same symmetry properties. An *ergodic spectrum* is defined by the property that its individual averages coincide with averages over the ensemble. Thus, a Hamiltonian matrix $\mathscr{H}$ with an ergodic spectrum can be substituted by a *random matrix* whose $K \times K$ elements are random variables. The method is familiar from its use in statistical mechanics, though the statistical natures of both theories differ. As Balian (1968) notes:

In statistical mechanics, some information is missing because the system is too complicated to be observed; the purpose is to make predictions in spite of this lack of knowledge. In the theory of random matrices, it happens that the Hamiltonian matrix $\mathscr{H}$ of the system is exactly known, but it is too complicated to be diagonalized practically. One deliberately discards some available information, assuming that the detailed form of $\mathscr{H}$ is irrelevant for the studied property, in order to satisfy a simplified (but statistical) mathematical treatment.

The presentation in this chapter is based on the review of Bohigas and Giannoni (1984) and on Balian (1968). The introductory review in Porter (1965) is an important background reference. That book contains the main papers on the subject up to 1965.

The main problem is that of determining the class of matrices to be admitted in the ensemble for a particular Hamiltonian and to define an appropriate measure for the ensemble. We know that Hamiltonians can always be represented by Hermitian matrices. If there exists an exact constant of the motion (such as angular momentum), the matrix can be separated into blocks. In this case the ensemble averages will refer to a single block. Underlying space–time symmetries impose restrictions on the ensemble. In the absence of spin, the matrix can be chosen to be real and symmetric, provided that the Hamiltonian is invariant with respect to time inversion. Otherwise (e.g., in the presence of a magnetic field), we must allow complex Hermitian matrices in the ensemble.

The measure $d\mathcal{H}$ in the space of matrices can be introduced in a natural way by the postulation of the *metric*,

$$ds^2 = \text{Tr}\,\delta\mathcal{H}\,\delta\mathcal{H}^\dagger \tag{8.29}$$

where $\mathcal{H}^\dagger$ is the Hermitian conjugate of $\mathcal{H}$. Hence, the 'distance' between two neighbouring matrices $\mathcal{H}$ and $\mathcal{H} + \delta\mathcal{H}$ can be written as the function of its elements,

$$ds^2 = \sum_{1 \le k \le K} (\delta H_{kk})^2 + 2 \sum_{1 \le j \le k \le K} (\delta H_{jk})^2, \tag{8.30}$$

in the case of real symmetric matrices, whereas for complex Hermitian matrices $\mathcal{H} = \mathcal{H}' + i\mathcal{H}''$, we have

$$ds^2 = \sum_{1 \le k \le K} (\delta H'_{kk})^2 + 2 \sum_{1 \le j \le k \le K} [(\delta H'_{jk})^2 + (\delta H''_{jk})^2]. \tag{8.31}$$

The introduction of a metric among N variables $\mathbf{x}$, in the form

$$ds^2 = \sum_{\mu,\nu}^{N} g_{\mu\nu}\delta x_\mu \delta x_\nu \tag{8.32}$$

implies the definition of the basic measure

$$dV = (\det g_{\mu\nu})^{1/2} \prod_{\mu=1}^{N} dx_\mu. \tag{8.33}$$

Therefore, the natural measure for the ensemble of real symmetric matrices is

$$d\mathcal{H} = 2^{K(K-1)/4} \prod_{1 \le k \le K} dH_{kk} \prod_{1 \le j \le k \le K} dH_{jk}, \tag{8.34}$$

and the measure in the space of complex Hermitian matrices has the form

$$d\mathscr{H} = 2^{K(K-1)/2} \prod_{1 \leq k \leq K} dH'_{kk} \prod_{1 \leq j \leq k \leq K} dH'_{jk} dH''_{jk}. \tag{8.35}$$

Any transformation for which the metric is invariant also preserves the measure $d\mathscr{H}$. The probability differential $P(\mathscr{H})d\mathscr{H}$ is also preserved; therefore, probability densities $P(\mathscr{H})$ are invariant with respect to real orthogonal transformations if $\mathscr{H}$ is real and symmetric, whereas for complex Hermitian matrices $P(\mathscr{H})$ will be invariant with respect to general unitary transformations. As a token of these invariance properties, we obtain the invariance of the *entropy*,

$$h\{P(\mathscr{H})\} \equiv -\int d\mathscr{H} \, P(\mathscr{H}) \log P(\mathscr{H}), \tag{8.36}$$

generalizing the definition (3.27) of entropy for a discrete probability distribution. The second and final postulate in the theory of random matrices is that $P(\mathscr{H})$ must be chosen so as to maximize the entropy. This is exactly the procedure in statistical mechanics – we presume the same probability for all accessible states; that is, in the absence of information, we choose the distribution with the maximum entropy.

Let us assume the knowledge of some average properties of the matrix $\mathscr{H}$:

$$\langle f_i \rangle \equiv \int d\mathscr{H} \, P(\mathscr{H}) f_i(\mathscr{H}) = C_i. \tag{8.37}$$

The distribution with maximum entropy with respect to any variation $\delta P(\mathscr{H})$, subject to the constraints (8.37), satisfies the equation

$$\int d\mathscr{H} \, \delta P(\mathscr{H}) \left\{ \log P(\mathscr{H}) - \sum_i \lambda_i f_i(\mathscr{H}) \right\} = 0, \tag{8.38}$$

since the integral over $\delta P(\mathscr{H})$ vanishes. Thus, the probability density is simply

$$P(\mathscr{H}) = \exp\left[\sum_i \lambda_i f_i(\mathscr{H}) \right], \tag{8.39}$$

and we determine the *Lagrange multipliers* λ_i by inserting (8.39) back into (8.37).

One way to guarantee that the eigenvalues of a finite matrix are bounded is to specify the norm

$$\langle \operatorname{Tr} \mathscr{H}\mathscr{H}^\dagger \rangle = \langle E_1^2 + E_2^2 + \cdots \rangle = C_2. \tag{8.40}$$

Combining this with the 'normalization constraint' ($\langle 1 \rangle = 1$), we obtain the probability density

$$P(\mathcal{H}) = \exp(\lambda_0 + \lambda_1 \operatorname{Tr} \mathcal{H} \mathcal{H}^\dagger) \qquad (8.41)$$

the *Gaussian orthogonal ensemble* (GOE) or the *Gaussian unitary ensemble* (GUE), depending on the symmetry properties of the matrix $\mathcal{H}$.

In the semiclassical limit we have to deal with an infinite matrix, with a mean level density that is known *a priori*, instead of a finite matrix with a finite norm. Thus, in this case we have a continuum of constraints

$$\langle d(E) \rangle = \langle \operatorname{Tr} \delta(E\mathbf{1} - \mathcal{H}) \rangle = \sum_j \delta(E - E_j) = \bar{d}(E), \qquad (8.42)$$

parametrized by the energy E. The resulting probability density for the *Weyl–Balian ensemble* is

$$P(\mathcal{H}) = \exp\left\{ \int dE\, \lambda(E) \operatorname{Tr} \delta(E\mathbf{1} - \mathcal{H}) \right\} = \exp\{\operatorname{Tr} \lambda(\mathcal{H})\}, \qquad (8.43)$$

or

$$P(\mathcal{H}) = \det \mu(\mathcal{H}), \qquad (8.44)$$

where

$$\mu(E) \equiv \exp\{\lambda(E)\}. \qquad (8.45)$$

The probability densities (8.41) and (8.43) depend only on the eigenvalues of $\mathcal{H}$. However, we need to know the measure $d\mathcal{H}$ in terms of the eigenvectors and eigenvalues of $\mathcal{H}$ before we can derive the combined probability density for the eigenvalues on their own. Let $\mathbf{E}$ be the eigenvalues of $\mathcal{H}$ and $\boldsymbol{\alpha}$ the parameters that specify the directions of its eigenvectors. The measure is then shown by Porter (1965) to have the form

$$d\mathcal{H} = \prod_{i<j} |E_i - E_j|^\beta \, d\mathbf{E}\, d\boldsymbol{\alpha}, \qquad (8.46)$$

where $\beta = 1$ for real matrices and $\beta = 2$ if they are complex. Integrating over all the directions of the eigenvectors, we thus obtain the combined eigenvalue probability densities as

$$P(\mathbf{E}) = C \prod_{i<j} |E_i - E_j|^\beta \exp\left\{ -(4\sigma^2)^{-1} \sum_k E_k^2 \right\} \qquad (8.47)$$

for the Gaussian ensembles and

$$P(\mathbf{E}) = \prod_{i<j} |E_i - E_j|^\beta \exp\left\{ \sum_k \lambda(E_k) \right\} \qquad (8.48)$$

for the Weyl–Balian ensemble.

In principle we can determine the function $\lambda(E)$ by integrating (8.48) over $K - 1$ variables and equating the result to $\bar{d}(E)$. A simpler alternative is to use the approximate method of Wigner (1957): For $P(\mathbf{E})$ we substitute $P[f(E)]$, where $f(E)\,dE$ is the number of eigenvalues in the interval dE. The allowed distributions $f(E)$ should be a set of K Dirac δ functions, but in the limit where $K \to \infty$ we can approximate $f(E)$ by a smooth function. The assumption is that $f = \bar{d}$ is the distribution that maximizes

$$\log P[f] \simeq \int dE\, f(E)\lambda(E) + (\beta/2) \int dE\,dE'\, f(E)f(E') \log|E - E'|. \quad (8.49)$$

[This neglects correlations in the second term of (8.47), which, it is hoped, are of short range.] Differentiating with respect to f, we obtain explicitly

$$\lambda(E) \simeq -(\beta/2) \int dE'\, \bar{d}(E') \log|E - E'|. \quad (8.50)$$

Exercise

Verify that this expression for λ makes $P(\mathbf{E})$ invariant with respect to changes in the energy scale.

In the case of the Gaussian ensembles, we have no *a priori* knowledge of $\langle d(E) \rangle$. Wigner's method can now be applied to (8.47), resulting in the integral equation

$$-(4\sigma^2)^{-1} E^2 + \beta \int dE' \langle d(E') \rangle \log|E - E'| = \text{const}, \quad (8.51)$$

with the solution

$$\langle d(E) \rangle = \begin{cases} (2\pi\sigma^2)^{-1}(4N\sigma^2 - E^2)^{1/2}, & |E| \leq 2N^{1/2}\sigma \\ 0, & |E| \geq 2N^{1/2}\sigma. \end{cases} \quad (8.52)$$

This expression for the mean level density coincides with the exact result for the Gaussian ensembles, obtained by direct integration. This fact reinforces our confidence as to the value of the approximation (8.50) for the Weyl–Balian density.

The semicircular form of the density of states for the Gaussian ensembles usually bears no relation to the true Weyl density. Even so, both distributions show *level repulsion* – $P(\mathbf{E}) \to 0$ as two levels approach each other. The similarity between the Weyl–Balian and the Gaussian distributions permits us to fit the latter to the former in a small energy range $dE \propto \hbar$, which is sufficient to obtain statistical properties of an individual semi-classical spectrum. The Gaussian distributions have two free parameters –

the width σ and the mean $\bar{E}$, taken as zero in (8.47). Thus, choosing

$$\bar{E} = E - 2\lambda/\lambda'(E), \qquad \sigma^2 = -\lambda/(\lambda')^2, \qquad (8.53)$$

where $\lambda' = d\lambda/dE$, we fit

$$\lambda(E) = -(4\sigma^2)^{-1}(E - \bar{E})^2 \qquad (8.54)$$

up to the first order in the Taylor expansion about any point inside δE. The only condition is that $\lambda(E) < 0$, which implies that $\bar{d}(E)$ should not have any very narrow peak. But this event can always be avoided with a change in energy scale, since the Weyl–Balian distribution with $\lambda(E)$ given by (8.50) is scale invariant.

The great advantage of fitting the Gaussian distributions to the correct distributions for an ergodic semiclassical spectrum is that many important statistical properties of the former are known exactly. This is the case of the *two-level correlation function*

$$Y_2(E_1, E_2) \equiv -\frac{P(E_1, E_2) - P(E_1)P(E_2)}{P(E_1)P(E_2)} \qquad (8.55)$$

and of higher correlation functions (see Bohigas and Giannoni, 1984, for references). In the case of the Gaussian orthogonal ensemble, $\beta = 1$, we obtain the limits

$$Y_2(r) \xrightarrow[r \to 0]{} 1 - \tfrac{1}{6}\pi^2 r + \tfrac{1}{60}\pi^4 r^3 + \cdots, \qquad (8.56)$$

$$Y_2(r) \xrightarrow[r \to \infty]{} (\pi r)^{-2} - (1 + (\cos \pi r)^2)(\pi r)^{-4} + \cdots, \qquad (8.57)$$

where

$$r = |E_2 - E_1|\bar{d} \qquad (8.58)$$

in a region in which $\bar{d}(E) = KP(E)$ can be considered constant. The distribution for the nearest-neighbour spacings has the form

$$P(r) = (\beta\pi^2/6)r^\beta + \cdots. \qquad (8.59)$$

We are now in a position to ascertain whether classically ergodic systems have spectra that are also ergodic. The computational evidence that this is indeed the case has become substantial. For example, Bohigas, Giannoni and Schmit (1984) obtained the histogram for $P(r)$ in fig. 8.7 from the analysis of 740 levels of Sinai's billiard. Similar calculations for the stadium as well as other statistics of the spectrum are also well fitted by the GOE. At the start of this section we restricted this analysis to systems without any special symmetries; therefore, fig. 8.7 deals with the spectrum for only one-eighth of Sinai's billiard (the desymmetrized billiard).

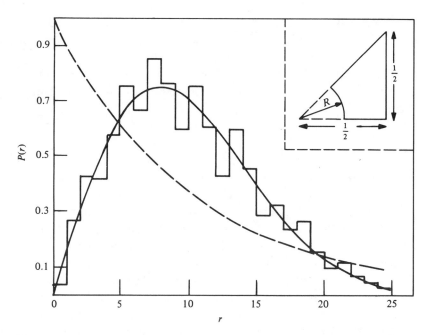

Fig. 8.7. Histogram of the nearest-neighbour spacings in the spectrum of Sinai's billiard. Dashed curve, Poisson distribution; solid curve, GOE; stepped curve, Sinai's billiard. (Courtesy of O. Bohigas, M. J. Giannoni and C. Schmit.)

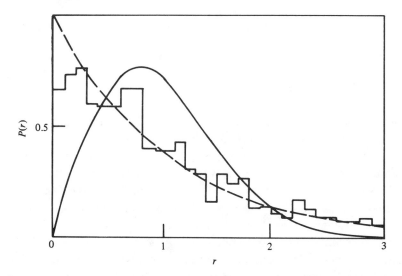

Fig. 8.8. Nearest-neighbour spacing for the quarter-circle billiard. Dashed curve, Poisson distribution; solid curve, GOE; stepped curve, quarter of a circle. (Courtesy of O. Bohigas, M. J. Giannoni and C. Schmit.)

As a comparison, fig. 8.8 presents the nearest-neighbour spacing of an integrable billiard – the quarter of a circle. There is no correlation between the way that different regions of the energy shell $H(\mathbf{I}) = E$ sweep through the 'dense' lattice of quantized tori in fig. 7.4. The fluctuations of the spectrum can thus be approximately modelled by the Poisson distribution. There is no level repulsion in this case.

8.5 Quantization of the cat map

Consideration of the discrete Poincaré map is always helpful in the study of continuous classical systems. However, there is no obvious way to quantize maps directly. The best available option is to consider the classical map as the stroboscopic map of a time-periodic Hamiltonian system and then deal with the *period operator* $\hat{U}_\tau$, that is, the evolution operator for the period τ of the Hamiltonian. This operator is unitary rather than Hermitian, but we can also investigate its eigenvalues and eigenfunctions. In particular we can verify whether its spectrum is ergodic, which in this instance means having the same fluctuation statistics as the *circular ensemble* of Dyson (1962) with properties similar to those of the Gaussian ensembles. The quantum map that has been most intensively studied is the 'standard map':

$$
\begin{aligned}
p' &= q - K \sin q, \\
q' &= q + p',
\end{aligned}
\tag{8.60}
$$

obtained from the periodic Hamiltonian,

$$
H(p, q, t) = \frac{p^2}{2} + \frac{K}{2\pi} \cos q \sum_n \delta(t - 2n\pi),
\tag{8.61}
$$

known as the *kicked rotor*, by Casati, Chirikov, Ford and Izrailev (1979).

Unfortunately the rigorous mathematical basis for quantum time-periodic systems is less developed than that for time-independent Hamiltonians. Indeed, one of the ways of obtaining such results, known as *Howland's method* (Howland, 1980), consists precisely of the elimination of time by expanding the phase space by two extra dimensions – just the reverse procedure of taking the Poincaré section!

An especially simple map to quantize is Arnold's cat map, presented in section 1.4 and used thereafter as a paradigm of classically chaotic motion. It has several unusual characteristics, but it is worthwhile to review briefly the results of Hannay and Berry (1980). In the case of the rotor (8.61), the variable q is an angle with period 2π (the phase space is a cylinder), so the conjugate variable – the angular momentum – is discretized. In the cat map, with $(p, q) = (\theta_2, \theta_1)$ in section 1.4, the momentum is also periodic (the phase space is a torus); therefore, the coordinates q are also discrete. The p and q wave functions for this system are necessarily 'combs' of equally spaced δ functions modulated by a (complex) periodic amplitude. The coordinate representation is

$$\langle q|\psi \rangle = \sum_{k=-\infty}^{\infty} \sum_{n=1}^{N} a_n \delta \left(q - \left(\frac{n}{N} + k \right) 2\pi \right), \qquad (8.62)$$

with the Fourier transform

$$\langle p|\psi \rangle = (2\pi\hbar)^{-1/2} \sum_{k=-\infty}^{\infty} \sum_{n=1}^{N} a_n \exp\left\{ i\hbar^{-1} 2\pi \left(\frac{n}{N} + k \right) p \right\}. \qquad (8.63)$$

The periodicity of the momentum must then be

$$\Delta p = N\hbar, \qquad (8.64)$$

so that the area of the torus is quantized as

$$\Delta p \, \Delta q = 2\pi\hbar N. \qquad (8.65)$$

In other words, the choice $\Delta q = \Delta p = 2\pi$ determines

$$\hbar = 2\pi/N, \qquad (8.66)$$

where N is the number of 'teeth' in one period of the wave function 'comb'. The semiclassical limit $\hbar \to 0$ here corresponds to $N \to \infty$. Use of the *Poisson identity*,

$$\sum_{j=-\infty}^{\infty} \delta(t - j) = \sum_{k=-\infty}^{\infty} \exp(i2\pi kt), \qquad (8.67)$$

leads to

$$\langle p|\psi \rangle = \sum_{j=-\infty}^{\infty} \left\{ \sum_{n=1}^{N} N^{-1/2} a_n \exp(inp) \right\} \delta \left(p - \frac{2\pi j}{N} \right), \qquad (8.68)$$

that is, a δ comb with N teeth in each period.

So far we have studied only the kinematics of the cat map. Its dynamics are based on the propagator corresponding to a linear canonical transformation, studied in section 7.6. The generating function

$$S(q, \mathsf{q}) = (\alpha \mathsf{q}^2 + 2\beta \mathsf{q}q + \gamma q^2)/2 \qquad (8.69)$$

defines the exact quantum propagator

$$\langle \mathsf{q}|\hat{U}|q \rangle = (-i\beta/2\pi\hbar)^{1/2} \exp\{i\hbar^{-1} S(\mathsf{q}, q)\}. \qquad (8.70)$$

The classical map has the form

$$
\begin{bmatrix} p \\ q \end{bmatrix} = \begin{bmatrix} a & d \\ b & c \end{bmatrix} \begin{bmatrix} p \\ q \end{bmatrix} = \begin{bmatrix} -\dfrac{\alpha}{\beta} & \dfrac{\beta^2 - \alpha\gamma}{\beta} \\[2mm] -\dfrac{1}{\beta} & -\dfrac{\gamma}{\beta} \end{bmatrix} \begin{bmatrix} p \\ q \end{bmatrix}.
\tag{8.71}
$$

Therefore, the propagator can be written as

$$
\langle q| \hat{U} |q \rangle = \frac{1}{2\pi} \left(\frac{iN}{b} \right)^{1/2} \exp\left\{ \frac{iN}{4\pi b} (aq^2 - 2qq + cq^2) \right\},
\tag{8.72}
$$

where the constants N, a, b, c are all integers.

In order to build up the torus propagator we now have to combine all the contributions of equivalent points $q + 2\pi m$ to the amplitude at q (fig. 8.9). The contributions of all equivalent points to the propagator at q will be out of phase and therefore cancel if q is not on the δ comb. Conversely points on the δ comb may receive all contributions in phase. The analysis becomes very simple in the special case where $b = 1$. Then

$$
\langle q| \hat{U} |q \rangle_{\text{torus}} = \frac{1}{2\pi} (iN)^{1/2} \sum_m \exp\left\{ \frac{iN}{4\pi} [aq^2 - 2q(q + 2\pi m) + c(q + 2\pi m)^2] \right\}
$$

$$
= \frac{1}{2\pi} (iN)^{1/2} \exp\left\{ \frac{iN}{4\pi} (aq^2 - 2qq + cq^2) \right\}
$$

$$
\times \sum_m \exp\left\{ i\pi [Ncm^2 + 2(cQ - \mathfrak{Q})m] \right\},
\tag{8.73}
$$

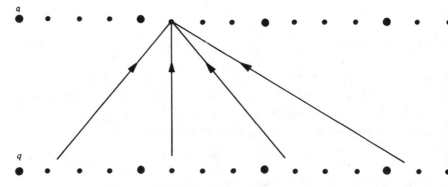

Fig. 8.9. Periodicity of the torus demands that the contribution of equivalent points $q + 2\pi m$ be included in the propagator.

defining the integers

$$Q = Nq/2\pi. \tag{8.74}$$

If Nc is even, then the exponential in the sum over m is always unity, so the sum diverges. Otherwise, it vanishes since Nc, being odd, implies an equal number of $+1$ and -1 terms. Instead of dealing directly with Dirac δ functions (in the case where Nc is odd), it is simpler to work with the discrete propagator

$$\langle q|\hat{U}|q\rangle_D = \left(\frac{i}{N}\right)^{1/2} \exp\left\{\frac{iN}{4\pi}(aq^2 - 2qq + cq^2)\right\}, \tag{8.75}$$

with the normalization

$$\sum_q |\langle q|\hat{U}|q\rangle_D|^2 = 1. \tag{8.76}$$

The analysis becomes much more intricate if b is a general integer. In particular, it is not always true that the discrete propagator has the same quadratic exponent as the plane propagator used to construct it. Hannay and Berry (1980) obtain general conditions for quantization: The matrix for the classical map must have the checker board structure

$$\begin{bmatrix} \text{even} & \text{odd} \\ \text{odd} & \text{even} \end{bmatrix} \quad \text{or} \quad \begin{bmatrix} \text{odd} & \text{even} \\ \text{even} & \text{odd} \end{bmatrix}. \tag{8.77}$$

This restriction prevents the quantization of the cat map (1.61) itself, though we can quantize the hyperbolic map with the matrix

$$\begin{bmatrix} 2 & 3 \\ 1 & 2 \end{bmatrix}.$$

This has $b = 1$, so the discrete propagator is

$$\langle q|\hat{U}|q\rangle_D = \left(\frac{i}{N}\right)^{1/2} \exp\left\{\frac{iN}{2\pi}(q^2 + qq + q^2)\right\}. \tag{8.78}$$

The elliptic map

$$\begin{bmatrix} 0 & -1 \\ 1 & 0 \end{bmatrix}$$

also has $b = 1$, so

$$\langle q|\hat{U}|q\rangle_D = \left(\frac{i}{N}\right)^{1/2} \exp\left\{-\frac{iN}{2}qq'\right\}. \tag{8.79}$$

We achieve a better grasp of the quantum cat map from the Wigner function. It was found in section 7.4 that the Wigner function propagates classically in a plane phase space, under the action of a linear map. For $b = 1$ the quantum map has a discrete propagator with the same form as the plane propagator, so the torus Wigner function propagates a classical distribution. A small complication arises from the need generally to include half-integer multiples of $2\pi/N$ for both the p and the q coordinates, as discussed by Ozorio de Almeida (1984). However, Hannay and Berry (1980) show that these supplementary coordinates can be omitted if N is odd. In this case we can consider Wigner functions defined on the $N \times N$ points of the quantized torus and normalized so that

$$\sum_{p,q} W(p,q) = 1. \tag{8.80}$$

All the quantized coordinates are multiples of $2\pi/N$, so the effect of the (integer) linear map is merely to permute the discrete points of the quantized torus. Each of these points has a periodic orbit with a period $n_i \leq N^2$. After n iterations, where n is the minimum common multiple of the n_i, the map brings any Wigner function back to itself. In other words the quantum map is periodic with period $n(N)$. Figure 8.10 shows all the

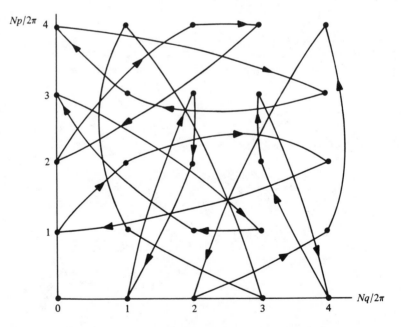

Fig. 8.10. All the cycles of the linear map (8.78) for the discretized torus with $N = 5$.

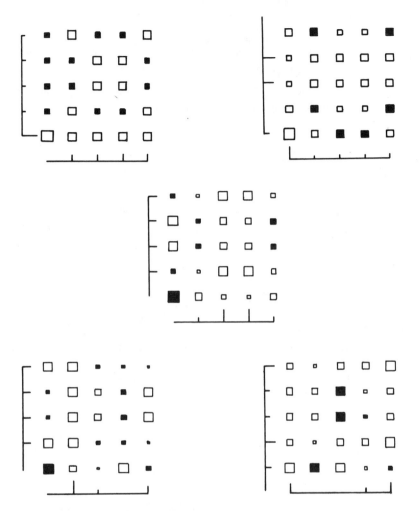

Fig. 8.11. Wigner functions for the same map as in fig. 8.10. The values of W are proportional to the size of the squares, and negative values appear as black squares. The lines to the left and underneath each figure are the projected probability densities. (Courtesy of J. H. Hannay and M. V. Berry.)

cycles of the map (8.78) for $N = 5$. An eigenstate of a quantized map has a Wigner function such that all points on the same periodic orbit have the same amplitude. Except for this restriction, the Wigner functions for the eigenstates of (8.78) bear no obvious relation to the classical periodic orbits. The five Wigner functions depicted in fig. 8.11 correspond to the

same quantum map whose cycles appear in fig. 8.10. In fig. 8.11 the values
of W are proportional to the size of the squares, and negative values
appear as black squares. The probability densities for the q and the p
representations appear, respectively, beneath and to the left of each figure.

The eigenvalues of the propagator fall on the unit circle. Since the
quantum map is periodic, the N eigenvalues have values

$$\alpha_j = k_j 2\pi/n. \tag{8.81}$$

Figure 8.12 shows the graph of $n(N)$ computed by Hannay and Berry
(1980) for the map (8.78). In contrast, the elliptic map (8.79) has $n = 4$ for
all $N \geq 4$. Thus, there is a marked difference between the spectra of maps
corresponding to classically integrable or chaotic motion. However,
though $n(N)$ is highly irregular, computations of $n(N)$ up to $N = 10^5$
indicate that

$$n(N) \sim CN, \tag{8.82}$$

with $C < 1$. The serious implication is that generally there are some
degenerate levels and there is no chance for there to be level repulsion in
the semiclassical limit as $N \to \infty$. This would be possible if $n(N) \propto N^2$, for

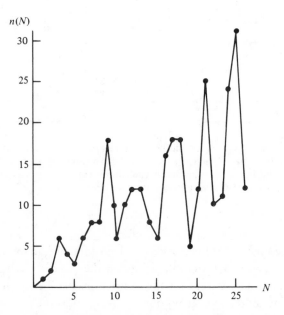

Fig. 8.12. The graph of $n(N)$. (Courtesy of J. H. Hannay and M. V. Berry.)

example. Though the cat map is a paradigm of classically chaotic motion, the quantum cat map does not have an ergodic spectrum. This is the only known counterexample of the rule presented in the previous section. The explanation seems to lie in the discretization of the quantum torus.

9

Periodic orbits in quantum mechanics

The quantization of classically integrable systems relies explicitly on the invariant tori that foliate phase space. No distinction is made in the traditional semiclassical theory as to whether the motion on a quantized torus is periodic or only quasi-periodic. A strong perturbation may break up tori with incommensurate frequencies as well as those with periodic orbits. The only surviving smooth invariant manifolds, beyond the energy shell itself, will then be isolated periodic orbits. The semiclassical limit of stationary states must therefore bear some relation to these invariant curves. The uncertainty principle prevents us from attempting to tighten too closely the association between a given orbit and a specific state. In the last resort a state has to rely on a collection of periodic orbits or at least on a periodic orbit dressed with the local motion around it.

The basic instrument for evoking the contribution of periodic orbits is the semiclassical propagator for points that return to their initial positions. The classical orbits determining this propagator are dominated by the periodic orbits, in a way that is explained in section 9.1. On the other hand, the propagator can be represented as a sum over the intensities of the stationary wave functions. A local analysis establishes that individual wave intensities may exhibit strong 'scars' along some periodic orbits. Taking the trace of the propagator, we obtain a formula for the density of states as a sum over all the periodic orbits.

Evaluating the trace of the propagator by the method of stationary phase, we obtain in section 9.3 the contribution of each periodic orbit together with the surrounding linearized motion. This works well for unstable periodic orbits and hence for ergodic systems. The stationary-phase contributions of stable periodic orbits suffer from a small-denominator problem, reminiscent of the classical theory presented in chapter 4. The solution to this problem is obtained in section 9.4 by substituting the resonant normal forms derived in section 4.4 for the inadequate linearization of the motion. Similar methods allow us to follow the periodic orbit sum through the transition of the breaking of integrability. In the opposite case of classical systems with positive entropy, the combination of the uniformity principle of section 3.7 with the periodic orbit sum leads to a partial dynamical explanation of ergodic spectra.

9.1 Stationary states and the propagator

The attempt to refine the classical limit of the Wigner function for an ergodic system by means of the pure-state condition in section 8.2 revealed a direct dependence on individual orbits in the energy shell. However, it is not easy to seek orbits with the property that their end points define a chord with a preestablished midpoint! It is more natural to return to the semiclassical propagator presented in section 7.6. Again we find a dependence on individual orbits, but these are defined directly by the variational principle for the action of paths between two points $\mathbf{q}$ and $\mathbf{q}$.

The propagation of a nonstationary semiclassical wave function is tied to the motion of a classical Lagrangian surface. Neither depends on the existence of invariant tori in the classical system. The specific case in which the Lagrangian surface starts out as the plane $\mathbf{q} = $ const. defines the semi-classical propagator. Alternatively, the propagator can also be envisaged as the coordinate representation $\langle \mathbf{q} | \hat{U} | \mathbf{q} \rangle$ of the evolution operator (7.86). This is a function of the Hamiltonian, so it has the diagonal energy representation

$$\hat{U}(t) = \sum_k |\psi_k\rangle\langle\psi_k| \exp(-i\hbar^{-1}E_k t). \tag{9.1}$$

Returning to the coordinate representation, we obtain

$$\langle \mathbf{q} | \hat{U} | \mathbf{q} \rangle = \sum_k \langle \mathbf{q} | \psi_k\rangle\langle\psi_k | \mathbf{q}\rangle \exp(-i\hbar^{-1}E_k t). \tag{9.2}$$

The crucial importance of this equation is that it ties the sum over unknown stationary states to a familiar semiclassical wave function.

We can easily invert (9.2) by taking its Fourier transform. However, we should proceed with some caution, because this would imply the knowledge of the semiclassical propagator even in the limits $t \to \pm\infty$. For a bound system, the originally plane Lagrangian surface will be infinitely convo-luted, in a manner analysed by Berry and Balazs (1979). The caustics then proliferate, destroying the simple semiclassical form of the propagator. For this reason it is safer to multiply the propagator by a hat function (unit near the origin while decaying to zero for $t \to \pm\infty$) before taking the Fourier transform. This is conveniently defined as the Fourier transform of a member of a family of peaked functions $\delta_\lambda(E)$ whose limit as $\lambda \to 0$ is the Dirac δ function:

$$\Delta_\lambda(t) = \int_{-\infty}^{\infty} dE\, \delta_\lambda(E) \exp(-i\hbar^{-1}tE), \qquad (9.3)$$

so that the width of the hat function $\Delta_\lambda(t)$ is $\hbar/\lambda$.

Convenient choices for $\delta_\lambda(E)$ are families of normalized Gaussians or Lorentzians. In the latter case Δ_λ is an exponential, so the multiplication of the hat function can be interpreted as the addition of an imaginary part to the energy. In the work of Balian and Bloch (1974) the semiclassical limit is indeed related to the analytic continuation of classical mechanics, but we shall deal here only with real systems, following Gutzwiller (1967, 1971).

The Fourier transform thus becomes

$$(2\pi\hbar)^{-1} \int_{-\infty}^{\infty} dt\, \Delta_\lambda(t) \langle \mathbf{q} | \hat{U} | \mathbf{q} \rangle \exp(i\hbar^{-1}tE)$$

$$= \sum_k \delta_\lambda(E - E_k) \langle \mathbf{q} | \psi_k \rangle \langle \psi_k | \mathbf{q} \rangle. \qquad (9.4)$$

For the choice $\lambda \sim \hbar$ the width of the hat does not diverge as $\hbar \to 0$. We can then safely use the semiclassical propagator, but this leads only to results for averages over a large number of states, since the averge energy level spacing is $(\bar{d})^{-1} = O(\hbar^L)$. To single out isolated states, we must take $\lambda = O(\hbar^L)$, which requires an extrapolation of the semiclassical propagator. We shall find that in some cases good results are obtained even in this limit.

Each term on the right side of (9.4) is the density matrix for a pure state $|\psi_k\rangle$. Taking the Weyl transform of the left side, we obtain a representation of the Wigner function in terms of the individual orbits in the energy shell, just as in section 8.2. It is the existence of this alternative path that permits us to avoid calculating the amplitude of the Wigner function through the iteration of the pure-state condition. Note, however, that this representation in terms of individual orbits exists even for integrable systems, as an alternative to the torus Wigner functions discussed in section 7.5. We shall study the relation between these alternative 'representations' in the limited context of periodic orbits and the energy spectrum in section 9.4.

Let us consider the case in which $\mathbf{q} = \mathbf{q}$; that is, we have wave intensities on the right side of (9.4). The propagator will then be a sum of terms – one for each orbit that leaves the point $\mathbf{q}$ at $t = 0$ and returns after a time t. Since we integrate over all time, we must also include the *zero orbit* – the orbit at any point $\mathbf{q}$ for $t = 0$. This contribution to the integral can be evaluated on its own, because all other orbits that return to $\mathbf{q}$ do so after a finite time. We found in section 7.6 that the propagator is singular as $t \to 0$, so we start from the momentum propagator $\langle \mathbf{q} | \hat{U} | \mathbf{p} \rangle$. For $\mathbf{q} \to \mathbf{q}$, we can consider $H(\mathbf{p}, \mathbf{q})$ to be a function only of the momentum; therefore,

$$
\begin{aligned}
\langle \mathbf{p} | \hat{U} | \mathbf{p} \rangle &\simeq \langle \mathbf{p} | \exp(-i\hbar^{-1} t H(\hat{\mathbf{p}}, \mathbf{q})) | \mathbf{p} \rangle \\
&= \exp\{-i\hbar^{-1} t H(\mathbf{p}, \mathbf{q})\}\, \delta(\mathbf{p} - \mathbf{p}).
\end{aligned}
\tag{9.5}
$$

The propagator is then the double Fourier transform

$$
\langle \mathbf{q} | \hat{U}(t \to 0) | \mathbf{q} \rangle = (2\pi\hbar)^{-L} \int d\mathbf{p}\, d\mathbf{p}\, \langle \mathbf{p} | \hat{U} | \mathbf{p} \rangle \exp\{i\hbar^{-1}(\mathbf{p} \cdot \mathbf{q} - \mathbf{p} \cdot \mathbf{q})\}
$$

$$
= (2\pi\hbar)^{-L} \int d\mathbf{p}\, \exp\{-i\hbar^{-1} t H(\mathbf{p}, \mathbf{q})\} \exp\{i\hbar^{-1} \mathbf{p} \cdot (\mathbf{q} - \mathbf{q})\}.
\tag{9.6}
$$

Introducing (9.6) into (9.4), after taking $\mathbf{q} = \mathbf{q}$, leads to the contribution of the zero orbits as

$$
(2\pi\hbar)^{-L} \int d\mathbf{p}\, \delta_\lambda(E - H(\mathbf{p}, \mathbf{q})).
\tag{9.7}
$$

We can eliminate all the other orbits that return to $\mathbf{q}$ by choosing $\hbar\lambda^{-1}$, the width of $\Delta_\lambda(t)$, smaller than the smallest time of return. In the semiclassical limit this choice is compatible with taking $\lambda \to 0$ in a classical energy scale; thus,

$$
(2\pi\hbar)^{-L} \int d\mathbf{p}\, \delta(E - H(\mathbf{p}, \mathbf{q})) = \sum_k \delta_\lambda(E - E_k) |\langle \mathbf{q} | \psi_k \rangle|^2.
\tag{9.8}
$$

We cannot neglect the width of the peaked function on the right side, though it is of $O(\hbar)$ since the mean level spacing is of $O(\hbar^L)$. However, if we do enforce the limit $\lambda \to 0$ on the right, there results an (unnormalized) wave intensity for the pure state $|\psi_k\rangle$, coincident with the projection of the energy shell Wigner function (8.1). We can therefore reinterpret this classical Wigner function as the correct semiclassical Wigner function, not for a pure state, but for the mixture of states in the classically narrow energy range λ. This result holds for integrable systems as well as ergodic systems.

Integrating (9.8) over the coordinate $\mathbf{q}$ results in

$$
(2\pi\hbar)^{-L} \int d\mathbf{p}\, d\mathbf{q}\, \delta(E - H(\mathbf{p}, \mathbf{q})) = \sum_k \delta_\lambda(E - E_k),
\tag{9.9}
$$

the Weyl rule for the density of states. In this form we can explicitly see how it describes a smoothed approximation to the level density. If we reduce the averaging width λ beyond $O(\hbar)$, the hat function in (9.4) will no longer avoid the contribution of the finite orbits returning to the point $\mathbf{q}$. These determine fluctuations around the Weyl rule, which will now be examined.

We can use the semiclassical propagators (7.85) directly in the integral (9.4). As the contributions of returning orbits appear only if the width of the hat function is $\hbar\lambda^{-1} \geq O(\hbar^0)$, we can evaluate the integral by stationary phase, considering $\Delta_\lambda(t)$ to be part of the smooth amplitude. The stationary time is determined by

$$E + \frac{\partial \sigma}{\partial t}(\mathbf{q}, \mathbf{q}, t) = E - H(\mathbf{p}(\mathbf{q}), \mathbf{q}) = 0. \tag{9.10}$$

The phase of the integral will be

$$Et_E + \sigma(\mathbf{q}, \mathbf{q}, t_E) = S_E(\mathbf{q}, \mathbf{q}), \tag{9.11}$$

where t_E is the solution of (9.10) and S_E is the reduced action, that is, the symplectic area subtended by the returning orbit. The contribution of each returning orbit to the wave intensity at $\mathbf{q}$ therefore has the form

$$C \left| \left(\frac{\partial^2 \sigma}{\partial t^2} \right)^{-1} \det \frac{\partial^2 \sigma}{\partial \mathbf{q} \, \partial \mathbf{q}} \right|^{1/2} \Delta_\lambda(t_E) \exp\{i\hbar^{-1} S_E(\mathbf{q}, \mathbf{q})\}. \tag{9.12}$$

There will be complex conjugate pairs of contributions with t_E and $-t_E$, since the integration extends between the limits $t \to \pm\infty$.

Many of these contributions oscillate rapidly with $\mathbf{q}$. Indeed, the phase of these oscillations is given by

$$\frac{dS_E}{d\mathbf{q}} = \frac{\partial S_E}{\partial \mathbf{q}}(\mathbf{q}, \mathbf{q}) - \frac{\partial S_E}{\partial \mathbf{q}}(\mathbf{q}, \mathbf{q}) = \mathbf{p}(\mathbf{q}) - \mathbf{p}(\mathbf{q}). \tag{9.13}$$

Thus, the condition for the phase of a returning orbit not to be highly oscillatory is that the returning momentum be identical to the initial momentum – only periodic orbits give stable contributions to the wave intensities!

The best manner of eliminating these rapid oscillations is to average the intensity over a region D^L, where $D \gg 2\pi\hbar|\mathbf{p} - \mathbf{p}|$ for nonperiodic orbits, as in section 8.1. There remains only the contribution of periodic orbits superimposed on the background Weyl intensity, due to the zero-period orbits. The variations of the averaged wave intensities in the neighbourhood of periodic orbits of chaotic systems were discovered by Heller (1984), who called them *scars*. If the energy width λ is chosen in such a way as to include only a single return of a periodic orbit, the intensity of its scar will vary sinusoidally with energy according to (9.12). From the relation

$$\partial S_E/\partial E = \tau_E \tag{9.14}$$

for the period of the orbit, we find that the wavelength of this oscillation is $2\pi\hbar/\tau_E$ – very long in comparison with the mean level spacing.

Integrating (9.12) over all $\mathbf{q}$, we obtain the contribution of all periodic orbits to the density of states up to a maximum period of $(\hbar\lambda^{-1})$. Again these are simple sinusoidal terms for each repetition of the orbit. The two following sections present the calculation of their amplitudes and the phase corrections due to the crossing of caustics. We will then be able to analyse the complete contribution of the multiple iterations of the same orbit. It is instructive to start with the study of simple systems with one freedom before proceeding to a general treatment.

9.2 The density of states for systems with one freedom

The closed level curves of the Hamiltonian of a bounded system with one freedom are themselves periodic orbits. A canonical coordinate transformation will not alter the semiclassical results, so it is convenient to use the coordinates (J, θ), introduced in section 2.5, where $2\pi J$ is the area inside a periodic orbit and

$$\dot{\theta} = \partial H/\partial J = \omega(J) \tag{9.15}$$

is a constant, since $H = H(J)$.

The complete action $\sigma(\Theta, \theta, t)$ supplies the usual relations

$$\partial\sigma/\partial\Theta = \mathfrak{J}, \qquad \partial\sigma/\partial\theta = -J. \tag{9.16}$$

On the other hand, integrating (9.15), we have

$$\omega(J) = (\Theta - \theta)/t. \tag{9.17}$$

Therefore

$$\left|\frac{\partial^2\sigma}{\partial\Theta\,\partial\theta}\right| = \frac{1}{t}\frac{dJ}{d\omega}. \tag{9.18}$$

Combining this result with

$$\frac{\partial^2\sigma}{\partial t^2} = -\frac{\partial E}{\partial J}\frac{\partial J}{\partial t} = \omega\frac{\partial J}{\partial\omega}\frac{\Theta - \theta}{t^2} \tag{9.19}$$

for m complete repetitions of the periodic orbit ($\Theta - \theta = m_2\pi$ and $t = m\tau$), we obtain the amplitude in (9.12) as

$$\left|\frac{\partial^2\sigma}{\partial q\,\partial q}\left(\frac{\partial^2\sigma}{\partial t^2}\right)^{-1}\right|^{1/2} = \frac{\tau}{2\pi}. \tag{9.20}$$

Thus, the contributions of each return of the periodic orbit have the same amplitude, in the case of a single freedom. If there is a single orbit with a given energy, the density of states will depend on the phase difference for successive returns of this orbit. It then becomes important to determine the constant in equation (9.12), or at least its phase. Provided that the Hamiltonian has the form $p^2/2m + V(q)$, the phase correction for the propagator will be $-\pi/4 - \mu\pi/2$, where μ is the number of caustics crossed by the orbit, according to the discussion in section 7.6. The integral over t furnishes a further $\pi/4$ times the sign of $\partial^2\sigma/\partial t^2$, that is, the sign of $\partial\omega/\partial J$. It is now important to consider carefully the change of variables $(p, q) \rightarrow (J, \theta)$. The problem is that this transformation alters the topology of the phase space – in the new variables there is no focal point where $\partial\Theta/\partial J = 0$, but in the Cartesian coordinates, the return of two neighbouring orbits separated by δE entails the crossing of at least one focal point, where $\partial\Theta/\partial p = \partial\Theta/\partial J = 0$ (fig. 9.1). There will be only one point if $\partial\omega/\partial J < 0$, whereas $\partial\omega/\partial J > 0$ corresponds to the existence of a pair of focal points for each winding of the orbit. Hence, the total phase correction for m repetitions is $-m\pi$, whatever the sign of $\partial\omega/\partial J$.

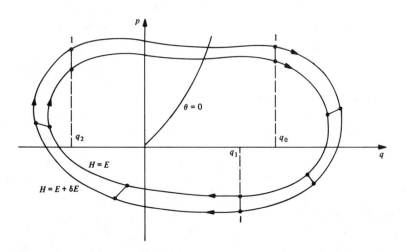

Fig. 9.1. The return of two neighbouring orbits separated by δE entails the crossing of at least one focal point. The orbit starting at q_0 with energy E has focal points at q_1 and q_2.

The contribution to the density of states of the mth return of the periodic orbit with energy E is the result of integrating (9.12) along the orbit. Since the integrand is independent of θ, we obtain

$$(2\pi\hbar)^{-1}\tau_E \exp(i\hbar^{-1}mS_E - m\pi)\, \Delta_\lambda(m\tau_E), \qquad (9.21)$$

where m ranges between $-\infty$ and $+\infty$. We can also include $m = 0$ – this is just the Weyl term, as can be seen by integrating the left side of (9.9) and then using (9.14). Taking then the limit $\lambda \to 0$ and summing over all the repetitions with the help of the Poisson identity (8.67) results in the density of states formula

$$d(E) = \hbar^{-1}\tau_E \sum_n \delta(\hbar^{-1}S_E - (2n+1)\pi) = \sum_n \delta(E - E_n). \qquad (9.22)$$

The equivalence between the periodic orbit formalism and Bohr–Sommerfeld quantization for one freedom was proved by Berry and Mount (1972) from the *Green function*

$$G(\mathbf{q}, \mathbf{q}, E) = (i\hbar)^{-1} \int_0^\infty dt \exp\{i\hbar^{-1}(E + i\varepsilon)t\} \langle \mathbf{q}|\hat{U}(t)|\mathbf{q}\rangle, \qquad (9.23)$$

where ε is a positive infinitesimal number. The equality

$$(-\pi)^{-1}\operatorname{Im} G(\mathbf{q}, \mathbf{q}, E) = \sum_n \langle \mathbf{q}|\psi_n\rangle\langle\psi_n|\mathbf{q}\rangle\delta(E - E_n) \qquad (9.24)$$

can be used as an alternative basis for the treatment of scars and the spectrum. I have avoided this intermediate step, because the propagator seems more intuitively accessible. Furthermore, there is the problem of the caustics. For one freedom there is actually an advantage to using the Green function – its caustics coincide with the turning points of the periodic orbit rather than the more subtle focal points of the propagator. However, in the general case the phase corrections are supplied directly by the Morse index, whereas the calculation of the phase of the Green function can be much more complicated. This subject is discussed by Möhring, Levit and Smilansky (1980).

9.3 Contribution of stable and unstable orbits

Periodic orbits of a system with two or more freedoms appear typically in one-parameter families. For an integrable system there will be $L-1$ further parameters, since we then have one-parameter families of periodic tori, but in this section we shall consider only periodic orbits that are isolated on each energy shell. The one-parameter family forms a two-dimensional annulus γ in phase space. It was shown in section 2.5 that we can make a coordinate transformation $(\mathbf{p}, \mathbf{q}) \rightarrow (J, \theta, \mathbf{P}, \mathbf{Q})$ to new canonical coordinates based on the annulus, such that $2\pi J$ is the symplectic area for each orbit and $\dot{\theta} = 2\pi/\tau$. As a consequence of this transformation

$$\frac{\partial H}{\partial \mathbf{P}}\bigg|_{\gamma} = \frac{\partial H}{\partial \mathbf{Q}}\bigg|_{\gamma} = 0 \tag{9.25}$$

and hence

$$\frac{d\omega}{d\mathbf{Q}}\bigg|_{\gamma} = 0. \tag{9.26}$$

Combining this equation with (9.16) and (9.17), which are still valid on γ (i.e., for $\mathbf{P} = \mathbf{Q} = 0$), we obtain

$$\frac{\partial^2 \sigma}{\partial \theta \, \partial \mathbf{Q}} = \frac{\partial J}{\partial \mathbf{Q}} = 0 \tag{9.27}$$

on each periodic orbit. The full action function for a return to $\mathbf{q}$ after m windings in the neighbourhood of a periodic orbit thus takes the form

$$\sigma_m(\mathbf{q}, \mathbf{q}) = \int_0^{2\pi m} J \, d\theta + S_m(\mathbf{Q}, \mathbf{Q}) - m\tau H. \tag{9.28}$$

The second term is a particular case ($\mathbf{\mathfrak{Q}} = \mathbf{Q}$) of the transverse action

$$S_m(\mathbf{\mathfrak{Q}}, \mathbf{Q}) = \int_{\mathbf{Q}}^{\mathbf{\mathfrak{Q}}} \mathbf{P} \cdot d\mathbf{Q} \tag{9.29}$$

taken along the orbit. This is not necessarily the generating function for the Poincaré map, since S_m defines a transformation for fixed time. However, the two actions can be identified by choosing the coordinates so that $\dot{\mathbf{Q}} = \dot{\mathbf{P}} = 0$ near the point $\mathbf{q}$. Along γ, the first term in (9.28) reduces to $2\pi m J$, but J varies along the orbit if $\mathbf{Q} \neq 0$.

The determinant of $\partial^2 \sigma/\partial \mathbf{q} \, \partial \mathbf{q}$ factorizes into a longitudinal term, calculated in the previous section, and a transverse term:

$$\left| \det \frac{\partial^2 \sigma}{\partial \mathbf{q} \partial \mathbf{q}} \left(\frac{\partial^2 \sigma}{\partial t^2} \right)^{-1} \right|^{1/2} = \frac{\tau_E}{2\pi} \left| \det \frac{\partial^2 S_m}{\partial \mathbf{\mathfrak{Q}} \, \partial \mathbf{Q}} \right|^{1/2}. \tag{9.30}$$

We can trivially integrate (9.12) over θ, so the contribution of a periodic orbit to the density of states has the form

$$(2\pi\hbar)^{-1} \tau_E A_m \exp\left\{ i(\hbar^{-1} m S_E - m\pi - \mu\pi/2) \right\} \Delta_\lambda(m\tau_E). \tag{9.31}$$

The amplitude A_m is defined as the trace of the unitary operator $\hat{U}_{\mathbf{P}}$

associated with the classical Poincaré map, as discussed in section 7.6:

$$A_m = \int dQ \langle Q | \hat{U}_P | Q \rangle$$

$$= (2\pi\hbar i)^{-(L-1)/2} \int dQ \left| \det \frac{\partial^2 S_m}{\partial \mathbb{Q} \, \partial Q} \right|^{1/2} \exp\{i\hbar^{-1} S_m(Q,Q)\}. \quad (9.32)$$

The phase correction in (9.31) has now been separated into two terms. The first, $m\pi$, has already been discussed in section 9.2 – each focal point in the γ annulus will be a general focal point, since $\partial^2\sigma/\partial q \, \partial q$ separates into two blocks. The other phase correction is determined by μ_m, the number of singular points of $\partial^2 S/\partial \mathbb{Q} \, \partial Q$ traversed by m windings of the periodic orbit.

By construction, the origin is a fixed point of the Poincaré map. Its generating function has no linear part, and the quadratic part defines the linearized map and the stability matrix $\mathcal{M}$. Henceforth, we shall deal only with the case where $L = 2$.

Exercise

Using (8.71), show that

$$\frac{d^2 S_m}{dQ^2}(Q,Q) = \det(\mathcal{M}^m - 1)\frac{\partial^2 S_m}{\partial \mathbb{Q} \, \partial Q}. \quad (9.33)$$

In the usual case where $(\partial \mathbb{Q}/\partial P)_Q^{-1} = (\partial^2 S/\partial \mathbb{Q} \, \partial Q) > 0$, the sign of the second derivative of S_m will be the opposite of $\det(\mathcal{M}^m - 1)$. Therefore, $d^2 S_m/dQ^2 < 0$, if $\mathcal{M}^m$ determines a direct hyperbolic map, whereas $d^2 S_m/dQ^2 > 0$ in the case of an elliptic map or of a hyperbolic map with a reflexion.

Evaluation of (9.32) by stationary phase is tantamount to approximating S_m by its quadratic terms; therefore,

$$A_m = \begin{bmatrix} 1 \\ -i \end{bmatrix} |\det(\mathcal{M}^m - 1)|^{-1/2}, \quad (9.34)$$

where the prefactor is only 1 in the case of a direct hyperbolic point. In this case the stability matrix will have the eigenvalues $(e^{m\alpha}, e^{-m\alpha})$, where α is the Lyapunov exponent, so the amplitude assumes the form

$$A_m = |2\sinh(m\alpha/2)|^{-1}, \quad (9.35)$$

first obtained by Gutzwiller (1971). The index $\mu_m = 0$, provided that the motion is hyperbolic along the entire orbit. For a hyperbolic map with reflection ($\mathrm{Tr}\,\mathscr{M} < -2$), the amplitude will be given by (9.35) if m is even because $\mathscr{M}^m$ will then determine a direct hyperbolic map. For m odd, the eigenvalues will be $(-e^{m\alpha}, -e^{-m\alpha})$ and therefore

$$A_m = \{2i\cosh(m\alpha/2)\}^{-1}. \tag{9.36}$$

The eigenvalues for a linearly stable map have the form $(e^{i\beta}, e^{-i\beta})$; hence,

$$A_m = -i|2\sin(m\beta/2)|^{-1}. \tag{9.37}$$

The amplitudes for the three cases above are not valid if α or β is very small. For the elliptic point we are also faced with a small-denominator problem in exact analogy to classical mechanics. The singularities in A_m arise for $\beta = 2n\pi$. Then the linearized Poincaré map for the nth return degenerates into the identity map – the sign of a periodic orbit resonance. The linear approximation thus surrounds the central periodic orbit with a continuum of periodic orbits, contradicting our assumption that the periodic orbit was isolated. It will be shown in section 9.4 that this difficulty can be surmounted by including the nonlinear terms of the resonant normal form studied in section 4.4.

Resonances are zeros of $\partial^2 S/\partial\mathfrak{Q}\,\partial\mathbf{Q}$ for $\mathfrak{Q} = \mathbf{Q}$. Similar zeros along the orbit lead to phase corrections of $\pi/2$. It will be shown at the end of this section that these corrections can be incorporated into a new amplitude

$$A_m^{-1} = 2i\sin(m\beta/2) \tag{9.38}$$

(with no modulus sign) within the periodic orbit contribution

$$(2\pi\hbar)^{-1}\tau_E A_m \exp\{im(\hbar^{-1}S_E - \pi)\}\,\Delta_\lambda(m\tau_E). \tag{9.39}$$

This formula differs from (9.31) by the absence of the explicit Morse phase due to transverse caustics. It is valid for any isolated orbit, inserting (9.38), (9.36) or (9.35) for A_m. The unidimensional phases $m\pi$, introduced by Miller (1975), must be kept.

For one freedom the periodic orbit is the energy shell itself; then the density of states depends only on successive returns of a single orbit. As the phase space dimensionality is increased, there usually appears on the energy shell an infinity of periodic orbits with arbitrarily long period. The sum of the successive repetitions of a single periodic orbit supplies the total intensity of its scar, which is just (9.39) divided by 2π. The density of states depends collectively on the interference between the total contribution of all the periodic orbits in the shell.

The total contribution of an unstable periodic orbit is absolutely convergent and tends to zero as $\alpha \to \infty$. In the particular case of free motion on a surface of constant negative curvature, the periodic orbit sum is actually exact – it corresponds to the *Selberg trace formula* (see the review article by Balazs and Voros, 1986). For $|m\alpha|$ large, we can approximate $2\sinh(|m\alpha|/2)$ by $\exp(|m\alpha|/2)$ and hence sum the resulting geometric series. Yet it is precisely the orbits with a small Lyapunov exponent that will have strong scars. So it is preferable to use

$$\frac{1}{2\sinh(|m\alpha|/2)} = \frac{\exp(-|m\alpha|/2)}{1-\exp(-|m\alpha|/2)} = \sum_{n=0}^{\infty} \exp\{-(n+\tfrac{1}{2})|m\alpha|\}. \quad (9.40)$$

It is now convenient to choose an exponential form for the hat function $\Delta_\lambda(t)$. The total intensity of a scar is then

$$\mathrm{Re}(2\pi^2\hbar)^{-1}\tau_E \sum_{n=0,m=1}^{\infty} \exp\{im[(\hbar^{-1}S_E-\pi)+i(n+\tfrac{1}{2})\alpha+i\hbar^{-1}\tau_E\lambda]\}. \quad (9.41)$$

Summing the geometric series in m, this becomes

$$\mathrm{Re}(4\pi^2\hbar)^{-1}\tau_E \sum_n (-1+i\cot\{2^{-1}[(\hbar^{-1}S_E-\pi)+i(n+\tfrac{1}{2}\alpha)+i\hbar^{-1}\tau_E\lambda]\}), \quad (9.42)$$

and using the expansion of the cotangent in partial fractions,

$$\cot z = \sum_{k=-\infty}^{\infty} (z-k\pi)^{-1}, \quad (9.43)$$

the intensity of a scar takes the final form

$$(4\pi^2\hbar)^{-1}\tau_E \sum_n \left\{-1+\sum_k \frac{2(n+\tfrac{1}{2})\alpha+2\hbar^{-1}\tau_E\lambda}{[\hbar^{-1}S_E-2(k+\tfrac{1}{2})\pi]^2+[(n+\tfrac{1}{2})\alpha+\hbar^{-1}\tau_E\lambda]^2}\right\}. \quad (9.44)$$

The $n=0$ term corresponds to a set of Lorentzian peaks in energy, first obtained by Heller (1984). The energy interval between the peaks is $2\pi\hbar/\tau_E$, using (9.14), and the width of each peak is $\lambda+\hbar(n+\tfrac{1}{2})\alpha/\tau_E$. Even in the limit $\lambda\to 0$, we obtain broad peaks – an unstable periodic orbit is not tied to a single eigenenergy. The width of the peak is of $O(\hbar)$, whereas the level spacing is of $O(\hbar^L)$. The distribution of the positive contributions to the scar intensity among the many states in an energy peak is still not known. Between the peaks the total contribution of the scars is negative, so that for these energies the orbits contribute weak *antiscars*. Thus, even partially smoothed wave intensities will show multiple oscillatons, corresponding to the scars and antiscars of the myriad periodic orbits, superimposed on the smooth Weyl distribution.

The terms with $n > 0$ in (9.44) bring quantitative corrections to the density of states. Further, broader Lorentzians with peaks at the same energies are added. For

$$(n + \tfrac{1}{2})\alpha > 2\pi \qquad (9.45)$$

the tails of the periodic Lorentzians will begin to superimpose, so that (9.44) loses its usefulness as a representation of the scar intensity, more easily obtained from (9.41). So if (9.45) is satisfied even with $n = 0$, the scar intensity is weak and sinusoidal. If (9.45) is satisfied only for $n = 0$, we obtain the Lorentzian profile with a weak sinusoidal correction (with the same maxima as the Lorentzians) due to the $n = 1$ term. For strong scars we have to add other Lorentzians up to the (9.45) limit. It follows that the energy width of a scar does not diminish by approaching a resonance with $\alpha \rightarrow 0$. Note that very close to this limit the linear approximation on which this whole discussion is based ultimately breaks down.

The intensity of a scar due to a stable periodic orbit can be treated in a similar manner, using the analytic continuation of (9.40) as proposed by Miller (1975). The orbit intensity is then given formally by

$$\mathrm{Re}(2\pi^2 \hbar)^{-1} \tau_E \sum_{n=0} \sum_{n=1} \exp\{im[\hbar^{-1} S_E - \pi - (n + \tfrac{1}{2})\beta + i\hbar^{-1} \tau_E \lambda]\},$$
$$(9.46)$$

which can be summed over m, yielding

$$(4\pi^2 \hbar)^{-1} \tau_E \sum_{n=0} \left\{ -1 + \sum_k \frac{2\hbar^{-1} \tau_E \lambda}{[\hbar^{-1} S_E - 2(k + \tfrac{1}{2})\pi - (n + \tfrac{1}{2})\beta]^2 + (\hbar^{-1} \tau_E \lambda)^2} \right\}$$
$$(9.47)$$

as the scar intensity. The density of states is obtained by multiplying (9.47) by 2π. This time we do obtain quantized energy levels as $\lambda \rightarrow 0$, determined by the conditions

$$S_E = \hbar\{2(k + \tfrac{1}{2})\pi + (n + \tfrac{1}{2})\beta\}. \qquad (9.48)$$

The convergence of (9.47) is just as tricky as the original sum, since the series in n cannot be truncated in this case. Even so the quantization condition (9.48) has a simple interpretation, based on the fact that the motion in the neighbourhood of a stable periodic orbit is at least quasi-integrable. It follows that the energy levels are supplied approximately by the quantization conditions of section 7.3. The orbit is surrounded by thin tori, with action variables J and I, respectively, along the orbit and transverse to it, that is $2\pi J = S_E$ for $I = 0$; a neighbouring torus will have a small value of I, and $2\pi J$ will be close to S_E. The quantization condition is that both I and J are a half-integer multiple of $\hbar$, so the periodic orbit itself is not directly tied to a quantum state. Given a torus with J correctly quantized, the action variable for the orbit with the same energy is approximately

$$J_{\text{orbit}} = (k + \tfrac{1}{2})\hbar + (n + \tfrac{1}{2})\hbar \frac{\partial H/\partial I}{\partial H/\partial J} \qquad (9.49)$$

for small n. This result agrees with (9.48) because

$$\frac{\partial H/\partial I}{\partial H/\partial J} = \frac{\dot\phi}{\dot\theta} = \frac{\beta}{2\pi}. \qquad (9.50)$$

The double infinity of levels given by (9.48) for each stable periodic orbit results from an unwarranted extrapolation of the linear approximation, valid only in the neighbourhood of the orbit. For n sufficiently large, the frequencies of the enveloping tori are no longer constant, and

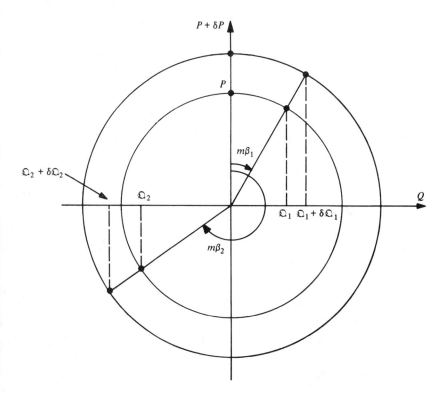

Fig. 9.2. If $(\partial\Omega/\partial P)_Q > 0$ as $m\beta \to 0^+$, then $(\partial\Omega/\partial P)_Q < 0$ for $m\beta > \pi$.

farther out they may cease to exist. Even so it is encouraging that we return approximately to valid quantization conditions by formally resumming the contributions of isolated periodic orbits.

To close this section, it remains only to justify the simplification achieved by Gutzwiller (1971) of the contribution of the mth return of a periodic orbit into (9.38) and (9.39), in spite of the transverse caustics. We obtained (9.37) under the assumption that $(\partial \mathcal{Q}/\partial P)_Q > 0$. However, fig. 9.2 shows that if the direction of increasing stability angle $m\beta$ is chosen to confirm this hypothesis as $m\beta \to 0^+$, it is falsified for $m\beta > \pi$. But for $m\beta = \pi$, $\partial^2 S/\partial \mathcal{Q} \, \partial Q = 0$, so we have the first focal point. The consequences of $m\beta$ passing through π are then that the prefactor in (9.34) changes from $-i$ to 1, while we must add the first Morse phase correction of $-\pi/2$. When $m\beta$ passes through 2π, $\partial \mathcal{Q}/\partial P$ becomes positive again, but now the Morse index is 2. Thus, the total contribution never changes phase at the odd-numbered caustics, whereas the phase change at the even caustics is π.

This result, incorporated into (9.38) and (9.39), is in agreement with the general rule that phase changes always accompany traversals through singularities. It is only the even caustics that lead to singular contributions to the density of states. The linearized approximation blows up there because it generates a continuum of periodic orbits with which it is unable to cope.

9.4 Amplitude of nonisolated orbits

The foregoing analysis holds only for isolated fixed points of the mth iteration of the Poincaré map. This fact is made evident by the alternative form of the amplitude (9.32):

$$A_m = \int d\mathbf{P} \, d\mathbf{Q} \langle \mathbf{Q} | \hat{U}_\mathrm{P} | \mathbf{P} \rangle \langle \mathbf{P} | \mathbf{Q} \rangle$$

$$= (2\pi\hbar)^{-(L-1)} \int d\mathbf{P} \, d\mathbf{Q} \left| \det \frac{\partial^2 S_m}{\partial \mathbf{Q} \, \partial \mathbf{P}} \right|^{1/2} \exp\{-i\hbar^{-1}[S_m(\mathbf{Q}, \mathbf{P}) - \mathbf{P} \cdot \mathbf{Q}]\}.$$

$$(9.51)$$

In the case that $\hat{U}_\mathrm{P}$ is the identity, the semiclassical form for $\langle \mathbf{\mathcal{Q}} | \hat{U}_\mathrm{P} | \mathbf{P} \rangle$ is exact, since $S_m(\mathbf{\mathcal{Q}}, \mathbf{P}) = \mathbf{P} \cdot \mathbf{\mathcal{Q}}$. Near the identity, this representation of the evolution operator has no caustics (hence, the real amplitude). The great advantage is that the Poincaré map is here given implicitly by

$$\frac{\partial S_m}{\partial \mathbf{\mathcal{Q}}}(\mathbf{\mathcal{Q}}, \mathbf{P}) = \mathbf{\mathfrak{P}}, \qquad \frac{\partial S_m}{\partial \mathbf{P}}(\mathbf{\mathcal{Q}}, \mathbf{P}) = \mathbf{Q}, \qquad (9.52)$$

so the fixed points correspond to the points of stationary phase in (9.51). If the fixed points are separated by many wavelengths, we can evaluate (9.51) by stationary phase, recovering the results in section 9.3. Otherwise, we must keep the nonquadratic terms of the action S_m needed to describe correctly the fixed points, that is, the periodic orbits that contribute collectively to a single scar or to the density of states.

Exercise

Deduce (9.51) directly from the mixed representation of the evolution operator

$$\langle q, \mathfrak{Q}|\hat{U}(t)|q, \mathbf{P}\rangle = \int dq' \, d\mathbf{Q'} \langle q, \mathfrak{Q}|\hat{U}(t)|q', Q'\rangle \langle q', \mathbf{Q'}|q, \mathbf{P}\rangle. \quad (9.53)$$

In the semiclassical limit, only orbits very close to each other will give a collective contribution; this will happen near a bifurcation of an isolated periodic orbit. The resonant normal forms for generic bifurcations of stable periodic orbits were studied in section 4.4. The complete classification by Meyer (1970) for systems with two freedoms is depicted in fig. 4.9. In principle, all that is needed is to substitute the quadratic approximation for S_m by the action corresponding to the appropriate normal form. In each case we have to deal with the mth iteration of the Poincaré map in the neighbourhood of a periodic orbit whose stability angle β is very close to $2\pi/m$. For $m = 1$ the integral (9.51) becomes the Airy integral, discussed in section 7.2. For $m = 2$ and 3 we obtain higher diffraction catastrophe integrals (the cusp A_2 for $m = 2$ and the elliptic umbilic D_4^- for $m = 3$) presented in section 6.3. For $m = 4$ the corresponding catastrophe is X_9, the first one with a continuous parameter in the classification of Arnold (1975). This is the parameter K in the function $x^4 + y^4 + Kx^2y^2$. Figure 4.9 shows both cases depending on whether K is greater or smaller than -2.

All the cases where $m \geq 5$ are better treated in polar coordinates, for which the amplitude becomes

$$A_m = (2\pi\hbar)^{-1} \int dI \, d\phi \left|\frac{\partial^2 S_m}{\partial I \, \partial\phi}\right|^{1/2} \exp\{i\hbar^{-1}[S_m(I, \phi) - I\phi]\}. \quad (9.54)$$

The normal form for the map $(I, \phi) \to (\mathfrak{I}, \Phi)$ has the generating function

$$S_m(I, \Phi) - I\Phi = \varepsilon I + cI^2 + aI^{m/2}\sin(m\Phi). \quad (9.55)$$

The resonance or bifurcation occurs for $\varepsilon = 0$ when satellite fixed points

Fig. 9.3. The action S_m has the shape of a Mexican hat. If the dimples along the rim are sufficiently small, the rim contributes as a full torus would rather than as a pair of orbits. If the whole hat is sufficiently small, the central periodic orbit cannot be separated from its satellites.

of the Poincaré section coalesce with the fixed point at the origin. Figure 9.3 shows the 'Mexican hat' shape of the action S_m, whose relative extrema at the centre and along the rim determine the periodic orbits. Inserting (9.55) into (9.54) and integrating over ϕ, we obtain

$$A_m = \hbar^{-1} \int_0^\infty dI\, J_0(\hbar^{-1} aI^{m/2}) \exp\{i\hbar^{-1}(\varepsilon I + cI^2)\}, \qquad (9.56)$$

where J_0 is the zeroth-order Bessel function.

The uniform approximation of this amplitude for all ε is furnished by Ozorio de Almeida and Hannay (1987). Here we shall discuss only the qualitative differences between the possible regimes. In the special case when $a = 0$, the Bessel function $J_0 = 1$ (the rim of the 'Mexican hat' is smooth). The integral then has a stationary phase for $I = I_0 = -\varepsilon/2c$, corresponding to the small torus of periodic orbits in the Birkhoff normal form. There are two regimes: If $\hbar^{-1}\varepsilon \gg 1$, the contribution of the torus and the central orbit can be separated. Alternatively, near a bifurcation ($\hbar\varepsilon \ll 1$) the torus and the central orbit contribute together. The important point is that these results are not fundamentally altered even if $a \neq 0$, as long as $\hbar^{-1}a \ll 1$. Classically the torus of periodic orbits is substituted by a single pair of stable and unstable orbits, no matter how small the constant a; but in the semiclassical limit $J_0 \to 1$ smoothly. We can thus describe the semiclassical amplitude in terms of a *quasi-torus* with action I_0. In the regime where $\hbar^{-1}a \gg 1$, we can approximate

$$J_0(x) \xrightarrow[x \to \infty]{} \left(\frac{2}{\pi x}\right)^{1/2} \cos\left(x - \frac{\pi}{4}\right), \qquad (9.57)$$

according to Abramowitz and Stegun (1964). Inserting this into (9.56), there results a pair of oscillatory integrals.

Exercise

Show that both integrals, approximated by the stationary-phase method, provide amplitudes corresponding to the satellite orbits. [The contribution of the end point $I = 0$ in (9.56) corresponds to the central stable orbit.]

One can readily adapt methods described above to the transition from an integrable to a quasi-integrable system. The main point is to note that the amplitude in section 9.2 and 9.3 was already defined in terms of the partial action-angle variables (J, θ) for a family of periodic orbits. Consider now an integrable system with Hamiltonian $H(\mathfrak{J})$. It is not usually true that the periodic orbits of a given torus with commensurate frequencies have the same topology as either of the two irreducible circuits, defining the actions $\mathfrak{J}_1$ and $\mathfrak{J}_2$. However, there always exists a canonical transformation (presented in section 6.4) to new action-angle variables, such that one pair becomes precisely the variables (J, θ) for the periodic orbit. The remaining conjugate pair (I, ϕ) determines the transverse motion. Inverting the equation $H(J, I) = E$, we obtain the reduced Hamiltonian $J(I; E)$. It was shown in section 2.5 that the orbits are then furnished by Hamilton's equations

$$\frac{dI}{d(-\theta)} = -\frac{\partial J}{\partial \phi}, \qquad \frac{d\phi}{d(-\theta)} = \frac{\partial J}{\partial I}. \tag{9.58}$$

Now the Poincaré map of the full system becomes just the stroboscopic map $(-\theta) = 2\pi$ of the reduced system. For an integrable system, J is independent of ϕ, so the flow reduces to a horizontal shear. The generating function for the corresponding stroboscopic map can then be obtained exactly:

$$m2\pi J(I; E) = S_m(I, \phi) - I\phi. \tag{9.59}$$

The right side of (9.59) is just the phase that has to be inserted in (9.54) for the amplitude to be evaluated. The left side is independent of ϕ, so we can immediately integrate over this variable, obtaining an amplitude like (9.56) with $a = 0$, that is, the amplitude for the Birkhoff normal form near an isolated periodic orbit. This is just as expected, since the Birkhoff normal form is an example of an integrable system. There is, however, a detail that should not be overlooked: In the previous treatment, the overall phase of the contribution was that of the central periodic orbit, even if the stationary torus were to contribute separately, since the torus was assumed to be close to the central orbit. In general it is preferable that the overall phase refer to the periodic orbits in the torus with action I_0.

It is therefore necessary to subtract the difference in phase between the origin and the torus, which is exactly $m2\pi J(I_0)$. The torus amplitude is thus

$$A_m = \hbar^{-1} \int_0^\infty dI \exp\{i\hbar^{-1} m2\pi [J(I) - J(I_0)]\}$$

$$= 2\pi \left| m\hbar \frac{\partial^2 J}{\partial I^2} (I_0) \right|^{-1/2} \exp(iv\pi/4), \qquad (9.60)$$

where v is the sign of the second derivative. It is important to note that the torus amplitude is of $O(\hbar^{-1/2})$ and therefore larger than that of an isolated periodic orbit.

The amplitudes of periodic tori in the density of states formula for an integrable system were first obtained by Berry and Tabor (1976). The method they used was the inverse of the transformation to the torus quantization conditions, which we made in section 9.2 for the particular case of one freedom.

In the case of a quasi-integrable system with the Hamiltonian $H = H_0(J, I) + \varepsilon H_1(J, I, \theta, \phi)$, we can also define the reduced Hamiltonian

$$J = J_0(I; E) + \varepsilon J_1(I, \phi; -\theta, E). \qquad (9.61)$$

However, the relation between the reduced Hamiltonian and the generating function is no longer given by (9.59).

Exercise

Show that up to first order in the perturbation parameter the generating function S_m for the Poincaré map is given by

$$m2\pi \bar{J} \equiv \int_0^{m2\pi} J(I, \phi + \omega_0 \theta; -m2\pi + \theta, E) d\theta = S_m(I, \phi) - I\phi, \quad (9.62)$$

where $\omega_0 = dJ_0/dI$. Near the periodic torus, this definition of $\bar{J}$ is equivalent to that obtained by first averaging the full Hamiltonian H over the fast variable θ as in section 6.5.

The perturbation adds a periodic term to the phase of the integrand of the amplitude formula. For a simple Chirikov resonance

$$\bar{J}(I, \phi) = J(I) + \varepsilon \cos \phi, \qquad (9.63)$$

we obtain the amplitude

$$A_m = 2\pi \left| m\hbar \frac{\partial^2 \bar{J}}{\partial I^2} (I_0) \right|^{-1/2} J_0(m2\pi\hbar^{-1}\varepsilon) \exp\left(\frac{iv\pi}{4}\right). \qquad (9.64)$$

Again, the asymptotic form (9.57) of the Bessel function, for sufficiently large perturbation parameters ε as compared with $\hbar$, results in distinct contributions for each isolated periodic orbit that survives the destruction of the torus I_0. Generally there will be an arbitrary periodic term $F(I, \phi)$ instead of the cosine in (9.63). Integrating first over the action I, we obtain

$$A_m = \int_0^{2\pi} d\phi \left| mh \frac{\partial^2 \bar{J}}{dI^2} \right|^{-1/2} \exp\left\{ i\hbar^{-1} m 2\pi [\bar{J}(I(\phi), \phi) - J(I_0)] + \frac{i\nu\pi}{4} \right\},$$

(9.65)

where $I(\phi)$ is defined by $\partial \bar{J}/\partial I = 0$. A uniform approximation to (9.65) is supplied by Ozorio de Almeida (1986). The contributions of the surviving periodic orbits will be isolated only if the phase difference between the maxima and minima of the exponent in (9.65) is much greater than 2π:

$$m\,\Delta \bar{J} \ll \hbar \qquad \text{(quasi-torus)}, \tag{9.66}$$

$$m\,\Delta \bar{J} \gg \hbar \qquad \text{(isolated orbits)}. \tag{9.67}$$

It is thus the difference in the symplectic area between a pair of periodic orbits, which survive the destruction of a periodic torus, that determines whether the orbits contribute separately or combine to form a quasi-torus.

The dependence on the number of repetitions in (9.66) and (9.67) seems to entail fundamentally different spectra for integrable systems and quasi-integrable systems – no matter how small the perturbation, the number of contributions satisfying (9.67) will always be infinite. It is therefore necessary to understand more fully the manner in which periodic orbits combine to manufacture the Bohr–Sommerfeld spectrum. Let us consider a quantized torus in an integrable system. The quantization condition, taking into account the Maslov indices, can be expressed in the following way: If all the independent circuits on this torus were periodic orbits, they would all supply zero-phase contributions to the density of states sum.

The quantized torus with frequencies ω is the limit of an infinite sequence of periodic tori, with frequencies that approximate ω arbitrarily well. Therefore, the periodic orbits of these tori have actions that are also good approximations of irreducible circuits on the quantized torus. The sum over infinite asymptotically coherent contributions results in the δ function of the density of states! Since none of the approximating tori has the exact quantized action, the difference between the quantized action and the approximate action of a given neighbouring torus grows proportionately to m. Thus, it is the first repetitions of the periodic tori that determine the quantization condition of an integrable system.

We conclude that a quasi-integrable system can be quantized by the Bohr–Sommerfeld rules of a neighbouring integrable system as long as (9.66) is satisfied with $m = 1$ for all orbits. This condition is probably still too restrictive for the local use of the torus quantization conditions near a surviving torus of a strong perturbation. The resonances near a surviving torus must become very weak the closer they are to the torus in the way described by Escande (1985). The implication of the foregoing reasoning appears to be that a quantized torus that survives a perturbation continues to contribute a δ function to the density of states, even if there are regions in phase space where the motion is strongly nonintegrable.

9.5 Correlation function for the density of states

Though of great theoretical interest, the periodic orbit sum formula in itself merely relates our ignorance of the density of states to the even more mysterious maze of periodic orbits. We can only attempt to explain the semiclassical fluctuations of the density of states by supplying further information about the distribution of periodic orbits. This is feasible for ergodic systems, because of the uniformity principle conjectured in section 3.6. An important consideration is that these (two-dimensional) systems have no stable periodic orbits: By virtue of the KAM theorem they would be surrounded by thin tori, which prevent most orbits from approaching the stable periodic orbit. Thus, there are no small denominators in the periodic orbit sum, the amplitudes being correctly given in the linearized approximation by (9.34).

The square of these amplitudes is closely related to the period distribution of periodic orbits (3.61) predicted by the uniformity principle, so we choose to work with the two-level correlation function for the fluctuations about the average density of states:

$$\langle \tilde{d}(E)\tilde{d}(E+\varepsilon)\rangle_E = \sum_{\substack{jm \\ j'm'}} (2\pi\hbar)^{-2}\langle \tau_j(E+\varepsilon)\tau_{j'}(E)A_{mj}(E+\varepsilon)A^*_{m'j'}(E)$$

$$\times \exp\{i[\hbar^{-1}mS_j(E+\varepsilon) - m\pi - \mu_{mj}\pi/2$$

$$- \hbar^{-1}m'S_{j'}(E) + m'\pi + \mu_{m'j'}\pi/2]$$

$$- \hbar^{-1}\lambda(|m\tau_j(E+\varepsilon)| + |m'\tau_j(E)|)\}\rangle_E, \tag{9.68}$$

where the fluctuation $\tilde{d}(E)$ is defined by (8.28). In the semiclassical limit we can expand $S_j(E+\varepsilon) = S_j(E) + \varepsilon\tau_j(E)$ in the exponent of (9.68), while neglecting ε in the slowly varying terms, as long as $\varepsilon \ll O(\hbar)$. Defining once more a peaked normalized function $\delta_\Lambda(E - E_0)$ centred on E_0 and of width Λ (independent of $\hbar$), we obtain

$$\langle \tilde{d}(E)\tilde{d}(E+\varepsilon)\rangle_E$$

$$= \sum_{\substack{jm \\ j'm'}} (2\pi\hbar)^{-2} \exp\left\{-i\left[(m-m')+\frac{(\mu_{mj}-\mu_{m'j'})}{2}\right]\pi\right\} \exp(i\hbar^{-1}\tau_j\varepsilon)$$

$$\times \int_{-\infty}^{\infty} dE\, \delta_\Lambda(E-E_0)\tau_j\tau_{j'}A_{mj}A_{m'j'}^* \exp\{-\hbar^{-1}\lambda(|m\tau_j|+|m'\tau_{j'}|)\}$$

$$\times \exp\{i\hbar^{-1}(mS_j-m'S_{j'})\}, \tag{9.69}$$

where the arguments of τ, A and S are always E. In general the last exponential will oscillate rapidly if $(m,j)\neq(m',j')$, unless $m\tau_j(E)=m'\tau_{j'}(E)$. This condition for accidental degeneracy of periods may certainly occur, but the evaluation of such terms by the stationary-phase method (considering δ_Λ to be a smooth amplitude) contributes only to $O(\hbar^{1/2})$, whereas the diagonal terms give contributions of $O(\hbar^0)$. So the correlation function is

$$\langle \tilde{d}(E)\tilde{d}(E+\varepsilon)\rangle_E$$

$$= (2\pi\hbar)^{-2} \sum_{jm} \langle \tau_j^2 |A_{mj}|^2 \exp\{\hbar^{-1}(i\varepsilon\tau_j - 2m\lambda|\tau_j|)\}\rangle_E + O(\hbar^{-3/2}). \tag{9.70}$$

We are now in a position to introduce the periodic orbit spectral function $f(\tau)$ given by (3.61) into (9.70), which becomes

$$\langle \tilde{d}(E)\tilde{d}(E+\varepsilon)\rangle_E = (2\pi\hbar)^{-2}\left\langle \int_{-\infty}^{\infty} d\tau |\tau| \exp\{\hbar^{-1}(i\varepsilon\tau - 2\lambda|\tau|)\} f(\tau)\right\rangle_E. \tag{9.71}$$

According to the uniformity principle, the asymptotic limit of $f(\tau)\to 1$ holds for all energies. Thus, the behaviour of the correlation function for small $\hbar^{-1}\varepsilon$ (so that its Fourier transform depends on the region where τ is large) will be

$$\langle \tilde{d}(E)\tilde{d}(E+\varepsilon)\rangle_E$$

$$\approx (2\pi\hbar)^{-2} 2\,\mathrm{Re} \int_0^{\infty} d\tau\, \tau \exp\{\hbar^{-1}(i\varepsilon - 2\lambda)\tau\} \xrightarrow[(\lambda/\varepsilon)\to 0]{} -(2\pi^2\varepsilon^2)^{-1}. \tag{9.72}$$

To compare this result with the statistical theory of ergodic spectra in section 8.4 we must first sort out the symmetry properties of the Hamiltonian. The deduction of (9.72) made no reference to symmetry, so this limit of the correlation function must be compared with that of the Gaussian unitary ensemble (GUE) with which it agrees. The Gaussian orthogonal ensemble (GOE) applies to ergodic systems with time reversal invariance: $H(-\mathbf{p}, \mathbf{q}) = H(\mathbf{p}, \mathbf{q})$. In this case periodic orbits come in symmetry pairs, except for the negligible proportion that are self-symmetric. Evidently, both orbits in a symmetry pair have exactly the same symplectic area, so there will be twice the usual number of terms for which $\exp\{i\hbar(mS_j - m'S_{j'})\} = 1$ in the sum (9.69). Since the amplitudes and the periods of a symmetry pair are also identical, we obtain

$$\langle \tilde{d}(E)\tilde{d}(E+\varepsilon) \rangle_E \xrightarrow[(\lambda/\varepsilon)\to 0]{} -(\pi^2\varepsilon^2)^{-1} \qquad \text{(GOE)}, \qquad (9.73)$$

in agreement with the dominant term of (8.57) – the ensemble average of $-\tilde{d}(E)\tilde{d}(E+\varepsilon)/(\tilde{d}(E))^2$. It may seem strange that the present result for small energy intervals should correspond to asymptotically large ε in section 8.4, but it must be remembered that $\varepsilon \ll O(\hbar)$ is not incompatible with $\varepsilon \gg O(\hbar^2) \sim 1/\tilde{d}$ in the semiclassical limit.

The fact that (9.73) agrees only with the asymptotic tail of the GOE correlation function is not necessarily the consequence of a deficiency of the periodic orbit sum for $\lambda \lesssim \tilde{d}$. The exact relation (for arbitrarily small λ) in the case of geodesic motion on a surface with constant negative curvature testifies to this end. There is a problem, however, with the neglected contributions of the accidental period degeneracies in (9.69). In the limit $(\lambda/\tilde{d}) \to 1$, the number of 'undamped' periodic orbits in the periodic sum increases to the extent that accidental period degeneracies may add a significant contribution.

In the original work on the uniformity principle, Hannay and Ozorio de Almeida (1984) deduced the correlation for the density of states in $\hbar^{-1}$, with fixed energy. Though somewhat unphysical, this has the advantage of dispensing with problems concerning semiclassical scales of the energy spectrum. In the case of billiards, analysed by Hannay (1984), the resulting spectrum is equivalent to the spectrum of the momentum (proportional to $E^{1/2}$). The first use of the uniformity principle directly in the energy spectrum was that by Berry (1985), who worked with the 'spectral rigidity' instead of the two-level correlation function.

It is worthwhile to conclude with a summary of our present partial knowledge concerning the energy spectra of chaotic systems. The common basis is the Weyl density. This large-scale average ($\Delta E \gg \hbar$) depends only on zero orbits; it is therefore insensitive to any dynamical feature of the system, except the relation between the volume and the energy of the energy shells. Though corrections are needed in the case of billiards, there is strong computational evidence that the Weyl density describes the real averaged spectra of integrable, ergodic and in-between systems. It appears that the spectra of almost all systems with positive entropy have ergodic fluctuations – their local averages ($\Delta E \ll \hbar$) can be calculated from appropriate ensemble averages. In the case of the two-level correlation function, we can successfully explain part of its form on the basis of the conjecture of the uniform distribution of periodic orbits. Obviously we need similar universal principles to explain further generic properties of the spectra of chaotic systems. So far, it is not clear how far these can be based on the uniformity principle itself, nor is it known whether *all* local statistical properties of the spectra of chaotic systems are ergodic. It has been computationally verified that the nearest-neighbour spacing (8.59) is usually ergodic in the case of positive entropy. The periodic orbit formalism is probably not the ideal probe for the study of such fine scales of the spectrum. It is in the intermediate range of averages ($\Delta E \sim \hbar$) that knowledge of a few short periodic orbits leads to a description of the nonuniversal spectral fluctuations.

Of course, there are many other important questions concerning the energy spectrum. Berry (1983) reviews the problem of eigenvalue degeneracies. Broadening our view, we find that comparatively little is known about the wave functions of chaotic systems. To some extent this also holds for quantum transport, diffusion and other nonstationary phenomena, though their absence from this book in no way reflects the rapid progress being achieved in their study.

References

Abramowitz, M. and Stegun, I. A. (1964). *Handbook of Mathematical functions.* Washington, DC: U.S. National Bureau of Standards.

Aguiar, M. A. M., Malta, C. P., Baranger, M. and Davies, K. T. R. (1987). *Ann. Phys. N.Y.* **180**, 167.

Arnold, V. I. (1963). *Russ. Math. Surv.* **18**, 85.

Arnold, V. I. (1965). *AMS Transl. Ser. 2*, **46**, 213.

Arnold, V. I. (1967). *Funct. Anal. Appl.* **1**, 1.

Arnold, V. I. (1973). *Ordinary Differential Equations.* Cambridge, MA: MIT Press.

Arnold, V. I. (1975). *Russ. Math. Surv.* **30**, 1.

Arnold, V. I. (1978). *Mathematical Methods of Classical Mechanics.* New York: Springer.

Arnold, V. I. (1982). *Geometrical Methods in the Theory of Ordinary Differential Equations.* New York: Springer.

Arnold, V. I. and Avez, A. (1968). *Ergodic Problems of Classical Mechanics.* Reading, MA: Benjamin.

Aubry, S. (1978). In *Solitons and Condensed Matter Physics,* ed. A. R. Bishop and T. Schneider. New York: Springer, 264.

Aubry, S. and Le Daeron, P. Y. (1983). *Physica* **8D**, 381.

Balazs, N. L. and Voros, A. (1986). *Phys. Rep.* **143**, 109.

Balian, R. (1968). *Nuov. Cim.* **57**, 183.

Balian, R. and Bloch, C. (1974). *Ann. Phys. N.Y.* **85**, 514.

Baltes, H. P. and Hilf, E. R. (1978). *Spectra of Finite Systems.* Mannheim: Wissenschaftsverlag.

Berry, M. V. (1976). *Adv. Phys.* **25**, 1.

Berry, M. V. (1977a). *Phil. Trans. Roy. Soc. Lon.* **287**, 237.

Berry, M. V. (1977b). *J. Phys.* **A10**, 2083.

Berry, M. V. (1978). In *Topics in Nonlinear Dynamics*, ed. S. Jorna. *Am. Inst. Conf. Proc.* **46**, 16.

Berry, M. V. (1981a). In *Singularities in Waves and Rays.* (Les Houches, Session 35), ed. R. Balian, M. Kleman and J. P. Poirier. Amsterdam: North Holland, 453.

Berry, M. V. (1981b). *Ann. Phys. N.Y.* **131**, 163.

Berry, M. V. (1983). In *Chaotic Behaviour of Deterministic Systems* (Les Houches, Session 36), ed. G. Ioss, R. G. H. Helleman and R. Stora. Amsterdam: North Holland, 171.

Berry, M. V. (1985). *Proc. Roy. Soc.* **A400**, 229.

Berry, M. V. and Balazs, N. L. (1979). *J. Phys.* **A12**, 625.

Berry, M. V., Hannay, J. H. and Ozorio de Almeida, A. M. (1983). *Physica* **8D**, 229.

Berry, M. V. and Mount, K. E. (1972). *Rep. Progr. Phys.* **35**, 315.

Berry, M. V. and Tabor, M. (1976) *Proc. Roy. Soc.* **A349**, 101.

Berry, M. V. and Upstill, C. (1980) *Progr. Opt.* **18**, 257.

Berry, M. V. and Wright, F. S. (1980). *J. Phys.* **A13**, 149.

Billingsley, P. (1965). *Ergodic Theory and Information.* New York: Wiley.

Birkhoff, G. D. (1927). *Acta Math.* **43**, 1.

Bohigas, O. and Giannoni, M. J. (1984). In *Mathematical and Computational Methods in Nuclear Physics* (Lecture Notes in Physics 209), ed. J. S. Dehesa, J. M. G. Gomez and A. Polls. New York: Springer, 1.

Bohigas, O., Giannoni, M. J. and Schmit, C. (1984). *Phys. Rev. Lett.* **52**, 1.

Bountis, T. C. (1981). *Physica* **3D**, 577.

Buminovitch, L. A. (1974). *Funct. Anal. Appl.* **8**, 254.

Casati, G., Chirikov, B. V., Ford, J. and Izrailev, F. M. (1979). *Stochastic Behaviour in Classical and Quantum Hamiltonian Systems* (Lecture Notes in Physics 93), ed. G. Casati and J. Ford. New York: Springer, 334.

Chirikov, B. V. (1979). *Phys. Rep.* **52**, 263.

Collet, P., Eckmann, J.-P. and Koch, H. (1981). *Physica* **3D**, 457.

da Silva Ritter, G. L., Ozorio de Almeida, A. M. and Douady, R. (1987). *Physica* **29D**, 181.

Devaney, R. (1979). *Ann. N. Y. Acad. Sci.* **316**, 108.

Dirac, P. A. M. (1930). *The Principles of Quantum Mechanics*. New York: Oxford University Press.

Dyson, F. J. (1962). *J. Math. Phys.* **3**, 140. Reprinted in C. E. Porter, ed. (1965). *Statistical Theories of Spectra: Fluctuations*. New York: Academic Press.

Engel, W. (1958). *Math. Ann.* **136**, 319.

Escande, D. F. (1985). *Phys. Rep.* **121**, 165.

Feigenbaum, M. (1978). *J. Stat. Phys.* **19**, 25.

Feigenbaum, M. (1979). *J. Stat. Phys.* **21**, 669.

Goldstein, H. (1980). *Classical mechanics*, 2nd ed. Reading, MA.: Addison-Wesley.

Greene, J. M. (1979). *J. Math. Phys.* **20**, 1183.

Greene, J. M., Mackay, R. S., Vivaldi, F. and Feigenbaum, M. J. (1981). *Physica* **3D**, 468.

Guckenheimer, J. and Holmes, P. (1983). *Nonlinear Oscillations, Dynamical Systems and Bifurcations of Vector Fields*. New York: Springer.

Gutzwiller, M. C. (1967). *J. Math. Phys.* **8**, 1979.

Gutzwiller, M. C. (1971). *J. Math. Phys.* **12**, 343.

Hannay, J. H. (1985). In *Chaotic Behaviour in Quantum Systems*, ed. G. Casati. New York: Plenum Press, 141.

Hannay, J. H. and Berry, M. V. (1980). *Physica* **1D**, 267.

Hannay, J. H. and Ozorio de Almeida, A. M. (1984). *J. Phys.* **A17**, 3429.

Heller, E. J. (1984). *Phys. Rev. Lett.* **53**, 1515.

Henon, M. (1969). *Q. Appl. Math.* **27**, 291.

Herman, M. R. (1979). *Publ. Math. IHES*, **49**, 5.

Howland, J. (1980). *Mathematical Methods and Applications of Scattering Theory* (Lecture Notes in Physics 130), ed. J. A. de Santo, A. W. Saenz and K. Zachary. New York: Springer.

Katok, A. (1982). *Ergod. Theor. Dynam. Syst.* **2**, 185.

Keller, J. B. (1958). *Ann. Phys. N.Y.* **4**, 180.

Khinchin, A. Ya. (1957). *Mathematical Foundations of Information Theory*. New York: Dover.

Khinchin, A. Ya. (1963). *Continued Fractions*. Groningen: Noordhoff.

Kolmogorov, A. N. (1954). *Dokl. Akad. Nauk.* **98**, 527.

Lanford, O. E., III (1983). In *Chaotic Behaviour of Deterministic Systems* (Les Houches, Session 36), ed. G. Ioss, R. G. H. Helleman and R. Stora. Amsterdam: North Holland, 6.

Lichtenberg, A. J. and Leiberman, M. A. (1983). *Regular and Stochastic Motion*. New York: Springer.

Mackay, R. S. (1983). *Physica* **7D**, 283.

Mackay, R. S., Meiss, J. D. and Percival, I. C. (1984). *Physica* **13D**, 55.

Mandelbort, B. (1982). *The Fractal Geometry of Nature*. San Francisco: Freeman.
Margulis, G. A. (1969). *FunctionalAnalizi Ego Prilozhen* **3**, 89.
Maslov, V. P. and Fedorink, M. V. (1981). *Semi-Classical Approximation in Quantum Mechanics*. Dordrecht: Reidel.
Mather, J. N. (1982). *Topology* **21**, 457.
McDonald, S. W. and Kaufman, A. N. (1979). *Phys. Rev. Lett.* **42**, 1189.
Meyer, K. R. (1970). *Trans. Am. Math. Soc.* **149**, 95.
Miller, W. H. (1975). *J. Chem. Phys.* **63**, 996.
Möhring, K., Levit, S. and Smilansky, U. (1980). *Ann. Phys. N.Y.* **127**, 198.
Moser, J. (1956). *Comm. Pure Appl. Math.* **9**, 673.
Moser, J. (1962). *J. Nachr. Akad. Wiss., Götingen, Math. Phys.* **K1**, 1.
Moser, J. (1966). *Ann. Scuola Normale Sup. Pisa Ser. 3* **20**, 499.
Moser, J. (1973). *Stable and Random Motions in Dynamical Systems*. Princeton, NJ: Princeton University Press.
Moyal, J. E. (1949). *Proc. Camb. Phil. Soc. Math. Phys. Sci.* **45**, 99.
Oseledec, V. I. (1968). *Trans. Moscow Math. Soc.* **19**, 197.
Ozorio de Almeida, A. M. (1982). *Physica* **110A**, 501.
Ozorio de Almeida, A. M. (1983). *Ann. Phys. N.Y.* **145**, 100.
Ozorio de Almeida, A. M. (1984). *Rev. Bras. Fis.* **14**, 62.
Ozorio de Almeida, A. M. (1986). In *Quantum Chaos and Statistical Nuclear Physics* (Lecture Notes in Physics 263), ed. T. Seligman. New York: Springer, 197.
Ozorio de Almeida, A. M., Coutinho, T. J. S. B. and da Silva Ritter, G. L. S. (1985). *Rev. Bras. Fis.* **15**, 60.
Ozorio de Almeida, A. M. and Hannay, J. H. (1982). *Ann. Phys. N.Y.* **138**, 115.
Ozorio de Almeida, A. M. and Hannay, J. H. (1987). *J. Phys.* **A20**, 5873.
Parry, W. (1984). *Ergod. Theor. Dynam. Syst.* **4**, 117.
Parry, W. and Pollicot, M. (1983). *Ann. Math.* **118**, 573.
Peixoto, M. C. and Peixoto, M. M. (1959). *Ann. Acad. Bras. Ci* **31**, 135.
Peixoto, M. M. (1962). *Topology* **1**, 101.
Percival, I. and Richards, D. (1982). *Introduction to Dynamics*. Cambridge University Press.
Pesin, Ya. (1977). *Russ. Math. Surv.* **32**, 54.
Poincaré, H. (1899). *Les methodes nouvelles de la mechanique celeste*, 3 vols. Paris: Gauthier-Villars.
Porter, C. E., ed. (1965). *Statistical Theories of Spectra: Fluctuations*. New York: Academic Press.
Poston, T. and Stewart, I. (1978). *Catastrophe Theory and Its Applications*. London: Pitman.
Ramaswamy, R. and Marcus, R. A. (1981). *J. Chem. Phys.* **74**, 1385.
Richens, P. J. and Berry, M. V. (1981). *Physica* **2D**, 495.
Sinai, Ya. G. (1959). *Dokl. Akad. Nauk.* **25**, 768.
Sinai, Ya. G. (1968). *Transl. A.M.S.* **73**, 227.
Sinai, Ya. G. (1970). *Russ. Math. Surv.* **25–1**, 137.
Van Vleck, J. H. (1928). *Proc. Natl. Acad. Sci. USA* **14**, 178.
Voros, A. (1979). In *Stochastic Behaviour in Classical and Quantum Hamiltonian Systems* (Lecture Notes in Physics 93), ed. G. Casati and J. Ford. New York: Springer, 326.
Wigner, E. P. (1932). *Phys. Rev.* **40**, 749.
Wigner, E. P. (1957). *Can. Math. Congr. Proc.* **174**. Reprinted in C. E. Porter, ed. (1965). *Statistical Theories of Spectra: Fluctuations*. New York: Academic Press.

Index